Exploring Complexity – Volume 6

DISRUPTED BALANCE

Society at Risk

Edited by

Jan Wouter Vasbinder
NTU, Singapore

NEW JERSEY · LONDON · SINGAPORE · BEIJING · SHANGHAI · HONG KONG · TAIPEI · CHENNAI · TOKYO

Published by

World Scientific Publishing Co. Pte. Ltd.
5 Toh Tuck Link, Singapore 596224
USA office: 27 Warren Street, Suite 401-402, Hackensack, NJ 07601
UK office: 57 Shelton Street, Covent Garden, London WC2H 9HE

Library of Congress Cataloging-in-Publication Data
Names: Vasbinder, Jan W., editor.
Title: Disrupted balance : society at risk / [edited by] Jan W Vasbinder, NTU, Singapore.
Description: New Jersey : World Scientific, [2018] | Series: Exploring complexity ; Volume 6
Identifiers: LCCN 2018019727 | ISBN 9789813239210 (hardcover : alk. paper)
Subjects: LCSH: Social problems--Asia, Southeastern. | Sustainable development--
    Asia, Southeastern. | Crisis management--Asia, Southeastern.
Classification: LCC HN690.8.A8 D57 2018 | DDC 306.0959--dc23
LC record available at https://lccn.loc.gov/2018019727

British Library Cataloguing-in-Publication Data
A catalogue record for this book is available from the British Library.

First published 2018 (Hardcover)
Reprinted 2018 (in paperback edition)
ISBN 978-981-3276-52-9 (pbk)

For any available supplementary material, please visit
https://www.worldscientific.com/worldscibooks/10.1142/10963#t=suppl

Printed in Singapore

CONTENTS

FOREWORD

Most of the time, discussion about risk generally focuses on how it is perceived, how it can be quantified and monetised, and what its specific characteristics are. Arguments about risk mostly deal with specific details such as tipping points, as well as climate change and its different consequences like droughts, floods or rises in sea level. We weigh the real and potential costs of mitigating risks or adapting to them. In the end, the monetary aspects prevail.

The conference that provided the material for this book was different. The theme was 'Disrupted Balance—Society at Risk', and the central questions that each of the twelve speakers were asked to address were:

What does it mean for society to be at risk?
What does risk mean for our societies?
What does risk mean for the people within these societies?

Rather than finances, we were more interested in the effects of disruptions on different balanced equilibria, the evolutionary processes and dynamics that influence society and the people living in it, as well as the resilience of our societies to the risks brought about by natural or anthropogenically induced disruptions.

The conference was organised by Para Limes at Nanyang Technological University, Singapore (NTU Singapore), a unit at a dynamic young university that explores big questions for humanity beyond the boundaries of disciplines, institutes or nationalities.

Para Limes brought together 12 speakers from eight different countries and five different continents. Each of these speakers is a world-class expert in an area where major disruptions threaten to put society at risk. Topics discussed included food, health, water, geohazards, governance, terrorism and cyber security, cities and low density human settlements, national security, finance and leadership.

The talks were designed to capture the attention of leaders who deal with risk not as an academic issue, but as something they have to handle as part of their responsibilities in society. These leaders include decision makers from government, civil society, industry, banks, insurance companies and non-governmental organisations. All these individuals need to anticipate eventualities and risks in their domains.

With this book, we aim to reach out to those leaders and other decision makers with a powerful message, brought forth by the speakers in a remarkably coherent way:

> Unless we do something, the world is heading for a major crash. We generally know what needs to be done and we have the science, technology and local knowledge to do it. To act adequately, science and technology must be brought to the service of humanity faster and more effectively.

NTU Singapore embraces this message. In the time since the conference, NTU Singapore's new President, Professor Subra Suresh, has announced that the University will establish the NTU Institute of Science and Technology for Humanity, which will tackle a number of the issues discussed here. Thus, this book is timely and contributes to the understanding of how academia can be of more direct service and higher positive impact to society and humanity as a whole.

Professor Alexander J.B. Zehnder

Chair of the Sustainable Earth Office
Nanyang Technological University, Singapore

INTRODUCTION

By Jan W. Vasbinder

Human society is no stranger to catastrophe. No nation exists whose history and character are untouched by floods, earthquakes, volcanic eruptions, climate fluctuations, famine, war or pandemics.

Throughout history, the impact of these disruptions has been felt mostly locally or regionally. But the world is changing—a ballooning population, intense global connectivity and the unquenchable thirst of human consumption have synergised to make disruptions more frequent, intense and global.

These disruptions threaten our most fundamental systems. Our food system, for example, is wholly dependent on agricultural conditions in a few 'breadbasket' regions and on the smooth functioning of an already tightly stretched and complex global transportation system.

Zooming out from our homes and cities to an atmospheric view of our globe, the synergy of growth, complexity and connectivity will also intensify the impact of human-induced climate change, heightening the risk of natural disasters, pandemics and disruptions to food and water supplies.

Exacerbating the challenge is the fact that governance is also at risk. Our institutions, whether local, national or international, are ill-equipped to cope with disruption—they do not know how to deal with the consequences of accelerating interconnectedness and complexity, nor do they know how to make full use of the growing body of scientific knowledge that would help them do so.

Taking the wheel

It was in light of this state of affairs that Para Limes at Nanyang Technological University, Singapore organised 'Disrupted Balance—

Society At Risk', a conference focused on how societies can achieve resilience in the face of inevitable and major disruptions.

The conference, held in December 2016, brought together 12 speakers—experts in their various fields—who were unanimous in the belief that unless we act soon, these collective global developments will eventually lead humanity to a crash, from which the road to recovery will be a difficult one.

Their views have been gathered into this book, which explores how we might soften the blows of disruption by bringing into sharper focus the actions that will make our societies more resilient in the face of sudden or unavoidable shocks.

The good news is that the crash does not have to happen. Humanity does possess the scientific, technological and social knowledge to not only survive, but also emerge from the tumult with more resilient and sustainable societies. The most urgent question, therefore, is how we can act on this knowledge.

Initiating the conversation

The first step is to kickstart a conversation, informed by science and data, about the level of security we want, and the level of risk we are willing to take. What parts of our lives are we ready to concede or defend, and what will it cost? We have the knowledge to inform this conversation—if only we would use it.

For example, decades of research in the geophysical sciences have taught us how seriously earthquakes, tsunamis and volcanic eruptions will damage our neighborhoods and infrastructure (see page 21, Gambling with Geohazards). Just as importantly, we have data telling us *how* natural hazards will strike, giving us mastery over the risks of building and planning our cities.

Disruption opens up space for hard conversations about risk. It unsticks the status quo, which is often half the battle. While unprecedented growth and urbanisation might super-size the impact of natural hazards, our need to build new spaces creates exciting opportunities to redesign our urban housing and infrastructure for resilience (see page 65, The Risks and Resilience of Cities).

Also in need of redesign is our broken financial system. Unrestrained financial activities have led to unsustainable consumption and huge imbalances in social income and wealth. To avoid another global financial crisis, we need to have a conversation around dealing with radical uncertainty, so that society can be served by finance instead of the other way around (see page 105, From Owners to Stewards).

Mustering the will to act

Although we possess the knowledge and technology to soften the impact of disruptions, applying these tools is not so much an intellectual issue as it is an emotional or cultural one—a question of mustering the will to act in a coordinated, strategic fashion.

Much of the vulnerability in our food system, for example, stems from habit. We know that we eat the wrong diet, that we produce too much of the same food at too low a price, and that disruption to our over-complex food systems hurts the poor and leads to civil breakdown. We need the will to use this knowledge to change our diets, reduce waste and redesign the food system to give our communities better food security (see page 29, Nine Meals from Anarchy).

Water is another example. While our blue planet still provides enough, demographic and climate changes will eventually shift water availability on a global scale (see page 81, Water: The Human Element). Failing to manage that shift will plunge humanity into disaster; yet most people in the developed world still take water access for granted. We need to put conversations about issues like water at the top of our agendas, as Singapore has done from its beginning and as wise governments have done throughout history.

In addition to food and water security, another area that requires concerted strategic actions and a 'whole-of-government' approach is national security. Here, complex geopolitical, social, technological and economic phenomena interact as a system of systems, thus necessitating the use of foresight tools to test assumptions and plan for the unexpected (see page 57, Thinking the Unthinkable).

Engaging communities

Also emotional is the spectre of disease. With emerging infectious diseases comes the threat of a pandemic that could kill a substantial portion of humanity and destroy basic security. This is again heightened by the synergies affecting our world—climate change, shifting demographics and hyper-connectivity. Yet our scientific knowledge can buy us time to counter or mitigate the impact of a pandemic (see page 73, Paradoxes and Paradigms in Parasitology).

An equally important component of a coordinated response to infectious disease is community engagement and education—we must integrate pathogen monitoring and surveillance activities on multiple scales, from citizens to world-renowned scientists to government policy makers.

Community engagement is also crucial in countering a social pathogen—that of terrorism. This scourge too is changing hosts and forms; it travels on waves of intolerance, a rising tide in many parts of the world. Our increasing knowledge of terrorism has foiled many attacks, but we need to understand the networks, backgrounds, methods and motives that lead organisations, small groups or individual people to commit acts of terror.

Importantly, our knowledge is broadening as well as deepening. Communities of practice are being set up, where branding experts, psychologists, security experts and youth workers are coming together to share ideas about how to stem the tide of terror (see page 49, Threats Seen and Unseen).

Removing blind spots in governance

Although the immense and unpredictable changes in our societies present new threats, governing institutions can draw lessons from centuries of history. Among these lessons, perhaps the foremost is that societies are embedded in an ecosystem on which they depend for their survival. The globe is riddled with the ruins of mighty societies that failed to respect or adapt to the tides of changing climates and social changes (see page 89, Sprawl at Risk).

Exacerbating the challenge is that planning for disruption plays into the traditional blind spots of our governing institutions: short-termism, vulnerability to randomness, and wilful blindness to problems in the 'too-hard' basket (see page 13, Hunting Black Swans and Taming Black Elephants). Leaders who dare to expect the unexpected and prepare the world for the unthinkable are in eternally short supply (see page 97, Thinking the Unthinkable V2).

Instead of building larger and more uniform institutions, we may need to tailor our systems of governance to scales at which we can agree on principles and accommodate diverse interests, so that we can build cooperation to mitigate catastrophic outcomes. Thus, an important part of planning for disruption may lie in understanding the limits of our ability to manage and control the world around us, and in re-designing our governance institutions accordingly (see page 39, Governance in a Time of Disruption).

A frank assessment

Today's disruptions are dynamic, interdependent and complex, with more moving parts to the puzzle than ever before. The 12 chapters in this book each help to define the shape and size of the puzzle pieces, so that we may build more cohesive and resilient systems to support our societies in the face of inevitable disruptions.

These frank assessments of the fragility and risk inherent in our societies may seem pessimistic, pointing to the dark clouds of possibility while ignoring examples of progress. But only an honest assessment of our situation will give us the knowledge, strength and will to set humanity on a more optimistic and resilient course.

HUNTING BLACK SWANS AND TAMING BLACK ELEPHANTS

The resilient way Singapore faces crises

By Peter Ho

How should we respond when our online, urbanised and intensely inter-connected lives make disruptions more frequent, amplified and unpredictable? How can governments deal with shocks that could spring from anywhere to disturb the flow of life?

A cyber attack, a new virus, civil unrest, or economic turbulence; each of these can originate on one side of the Earth, yet disrupt life on the other within days, hours or minutes. This is the challenge of our age: facing problems that we cannot predict, pre-empt or take flight from, much less fight.

Of black swans and black elephants

Not long ago, most disasters arrived without warning—storms, floods, earthquakes, volcanic eruptions. Today, modern science helps to predict natural disasters and other upheavals with increasing accuracy. Such events are no longer seen as 'black swans'—rare events that take people by surprise.

So why is it often the case that no one takes any precautionary measures? The answer lies in a failure all human beings are prone to:

our cognitive biases or blind spots. Present costs and benefits occupy our mind's centre stage and push future risks to the background.

Behavioural economists call the focus on the now 'present-biased preferences'. In his book *Collapse*, Jared Diamond calls it 'creeping normality', the tendency—much like the proverbial frog in boiling water—not to notice how things get just a little bit worse each year, until a catastrophic collapse occurs. Our tardy response to climate change is a case in point: governments are often catatonic in the face of problems whose consequences are expected to be felt only in the distant future.

This leads me to a new member of my menagerie, the 'black elephant': a cross between the 'black swan' and the 'elephant in the room'. The black elephant is a problem that is actually visible to everyone, but no one wants to deal with it, so we pretend it is not there. When the problem blows up, we all feign surprise and shock, behaving as if it were a black swan.

Two black elephants that cast shadows over many populations are the threats of pandemics and terrorism—subjects very much on people's minds, but considered too hard to deal with. Yet such crises have much to teach us about resilience and response in uncertainty.

A tale of two pandemics

In 2013, a small Ebola outbreak in Guinea ballooned within a year into an international health emergency, leaving over 10,000 people dead. Experts say the tragedy could have been nipped in the bud if appropriate and aggressive action had been taken at the start. Unfortunately, the human mind tends to underestimate both sudden crises and slow-burn issues alike. The result is organisational hesitation: until events reach crisis proportions, no one acts. This reluctance to act is an example of a black elephant.

Singapore provides another case study, but one that provides lessons in response and resilience. On 25 February 2003, the severe acute respiratory syndrome (SARS) virus entered Singapore through three women who had stayed at a Hong Kong hotel where a doctor from China was also lodging. He was afflicted with SARS, which is thought to have emerged in China in November 2002.

The virus spread with frightening speed through Singapore's hospital system, confounding our medical authorities. There were many unknowns: authorities did not know how the virus spread, or why it spread so aggressively. The fatality rate was shocking. By the time the SARS crisis was declared over in Singapore, 33 out of the 238 infected patients had died.

SARS was a black swan for Singapore, shocking and frightening Singaporeans. Indeed, "a crisis of fear" was how the then Prime Minister Goh Chok Tong described the situation to the BBC. Overnight, visitor arrivals plunged, the entire tourism industry ground to a halt and the economy shrank.

A resilient response

Like it or not, there will be more of such crises. We have to ask ourselves this key question: if disruption is inevitable, how can we develop a system that is resilient to such shocks?

One answer is to not to overemphasise efficiency at the cost of redundancy and resilience. In the case of SARS, a critical early decision was to designate the pandemic a national crisis, not just a public health problem. All the nation's resources could then be harnessed against SARS. The Singapore Armed Forces (SAF) put an entire army division at the disposal of the health authorities; the police did likewise. Within weeks, the Defence Science and Technology Agency (DSTA) and Defence Science Organisation (DSO) National Laboratories developed contact tracing and infrared fever screening systems, tools that have now been adopted globally.

This could not have been achieved if the government had been obsessed about efficiency and optimisation. On their own, efficiency and optimisation are of course important—provided that everything goes according to plan. But things rarely do. Nassim Nicholas Taleb, who popularised the phrase 'black swan', notes that when events disrupt overly optimised systems, "errors compound, multiply, swell, with an effect that only goes in one direction—the wrong direction". This is not an argument for establishing bloated bureaucracies, but rather about having strategic capacity. The SAF and its supporting organisations like DSTA and DSO are reserves of the nation or an insurance premium. Without that 'fat' in the system, it is doubtful that Singapore would have responded to the SARS crisis as effectively.

Clearing the fog of fear

Another important lesson from the SARS crisis was the benefit of going against the instinct to downplay the situation. During Singapore's SARS outbreak, the authorities took many unprecedented actions, including contact tracing, temperature screening, and home quarantine with electronic monitoring bracelets and security monitoring. Initially, these measures were predictably denounced in the western press as "draconian". But many of these measures were then quietly adopted by other cities afflicted by SARS.

In an April 2003 interview with the *BBC*, Goh said, "I'm being realistic because we do not quite know how this will develop. If it becomes a pandemic, then that's going to be a big problem for us. I'd rather be proactive and overreact a little so that we can get people who should quarantine themselves to stay at home."

In other words, when dealing with serious and uncertain disruptions, it is often better to overshoot rather than undershoot.

But another issue was at play: fear. Fear rears its head during any crisis. Spreading trusted information manages fear, but sometimes goes against our instincts. During the SARS outbreak, Singapore chose transparency. The government gave the World Health Organization (WHO) unfettered access to information, inviting its representatives to attend daily meetings chaired by the Director of

Medical Services where all the data and information collated from the previous 24 hours was presented.

The government also laid bare the uncertainties and risks during SARS, avoiding giving false assurances. This formed a sense of trust which became a deep source of national resilience during the crisis. In other cases, communication and trust have proven to be a weak point. Take for instance South Korea's experience with Middle East respiratory syndrome (MERS) in 2015, arguably an example of undershooting.

When MERS hit South Korea, the government was criticised for its slow response to the outbreak, and for stoking fears by failing to effectively communicate public health risks. The lack of information led to the spread of groundless rumours that only served to amplify unfounded fears in a vicious circle.

South Korea struggled with its response to MERS even though the disease was not unknown and already had global authorities on high alert because of its high morbidity rate. Pandemics in recent years, such as SARS, should also have provided lessons. Clearly, while learning from the past might seem like a simple idea, it is hard to put into practice.

The black elephant of terrorism

Terrorism provides further food for thought about black elephants and how governments should respond to them. In particular, the situation in France makes for an interesting case study.

No stranger to the threat of terrorism, French security authorities have in recent years been most concerned about what they call 'third-generation migrants', the grandchildren of migrants who have no memories of the hard life their grandparents left behind but who nonetheless feel alienated in their country of birth.

With the emergence of the Islamic State of Iraq and Syria (ISIS), the problem has come home to roost over the last two years in the form of tragic attacks, including the 2015 attack on the offices of satirical weekly *Charlie Hebdo* and the November 2015 shootings and suicide bombings that killed 130 people.

The French have shown remarkable resilience in the face of these attacks, embodied in the rallying cry of *"Je suis Charlie"*. Yet underneath the symbolic rhetoric, the fabric of society may be tearing. After the *Charlie Hebdo* attack, there were at least 160 attacks on Muslim people and their communities. This heightened tensions that stretched back to 2005, when riots followed the death of two French youths of Malian and Tunisian descent as they fled the police.

Longer-term issues are clearly at play here, including social and economic exclusion, racial discrimination and the capacity of the secular state to integrate cultural and ethnic diversity.

Political actors in France have vowed to fight the threat of terrorism, but have also stoked zeal through cultural controversies such as the public wearing of the veil and *burka*.

The question is whether France's response is merely reactive, or if it exhibits both resilience and antifragility. A resilient society returns to the state it was at before the disruption. In contrast, an antifragile society reaches a new state, almost like a muscle that, tested by stress, grows stronger. In the context of terrorism, an antifragile response might include the promotion of social integration.

Putting community first

Singapore has had its own brushes with terrorism, such as the discovery and detention of 13 members of the hitherto-unknown Jemaah Islamiyah (JI) terrorist network in 2001. Since then, Singapore has detained more than 66 people under the Internal Security Act for terror-related activities. Besides these measures, the government has taken care to ensure that the Muslim community remains integrated into broader society, rather than give cause for alienation.

The cornerstone of Singapore's counter-terrorism strategy is a community response plan. This enhances community vigilance, community cohesion and community resilience. Singapore has built networks of community leaders and influencers by forming the Inter-Racial and Religious Confidence Circle (IRCC). These leaders have helped strengthen the understanding and ties between different races and religions. Muslim leaders not only speak out against those who distort Islam, but also use the media, mosque and madrasah to assert mainstream Islamic values.

The IRCC also builds social resilience through outreach. It has organised conferences, forums, dialogues and briefings to educate the community about key Islamic concepts which have been misinterpreted and misrepresented by terrorist and extremist groups such as JI, Al-Qaeda and ISIS.

Ultimately, terrorism and disease outbreaks are both about diffusion—of ideologies and pathogens. Both can be black elephants, risks we put off preparing for or avoid talking about altogether. In addition to implementing specific measures to secure installations or quarantine patients, the softer aspects—the diffusion of trusted information and weaving a tough yet supple social fabric—are equally important. The underlying layer of trust and cohesion is what societies need to build in times of peace, even as they learn to hunt down black swans and tame the black elephants that will surely visit societies and countries from time to time.

Peter Ho is the Senior Advisor to the Centre for Strategic Futures, a Senior Fellow in the Civil Service College, an Adjunct Professor at the S. Rajaratnam School of International Studies, a Visiting Fellow at the Lee Kuan Yew School of Public Policy, and SR Nathan Fellow at the Institute of Policy Studies. He is also Chairman of the Urban Redevelopment Authority of Singapore and Chairman of the Social Science Research Council.

GAMBLING WITH GEOHAZARDS

Extending our horizons with science

By Kerry Sieh

Although we may seldom stop to think about it, each of us faces risks every day, making minute by minute decisions about which risks are acceptable and which are not. When we decide to cross a road, for example, we are taking a calculated risk based on assumptions of how other road users are likely to behave. The cost to us as an individual may be high—serious injury or even death—but because we face the risk of crossing roads so frequently, we seldom avoid looking both ways before confidently striding across.

Human beings usually deal very well with frequent hazards, coming up with behaviours that ensure our survival. Recognising the dangers on the roads, for example, we obey traffic rules for our own safety and for the safety of our fellow citizens.

However, it is much harder for us to comprehend low-likelihood, high-impact risks, catastrophic events that happen only once every few hundred years or more. As an earth scientist, I see it as my role to bring attention to these infrequent events, helping people see into the distant past and extending our history so that we can be aware of what can happen and how often it happens.

Resilience in response to risk

Even when hazards occur relatively frequently geologically, we tend to let our guard down within a generation or two and start gambling with them again. Having a geological understanding of our past instead helps us to be prepared for potentially catastrophic geohazards such as earthquakes and volcanic eruptions.

In the case of Indonesia's Mount Merapi, a 2010 eruption led to the loss of hundreds of lives because hazard maps were drawn based on records that only went back to 1865. In 2010, however, the incandescent hot gas clouds flowed further than their predecessors had in recorded history, going past the hazard zones and catching people unawares. These exceptions to historical limits are important to know and to understand because many people continue to live on the flanks of Merapi today, including the 2.4 million inhabitants of nearby Yogyakarta.

In other cases, countries have responded to hazards by building resilience to low-likelihood hazards into their communities, just as we have developed traffic rules to reduce the risk of accidents. Taiwan, for example, has some of the strongest building codes and enforcement mechanisms in the world to deal with their infrequent quakes. These have resulted in buildings so robust that when the Chichi earthquake struck in 1999, structures built right next to or even on the fault line remained standing.

As much as the resilience of the buildings during the Chichi earthquake are a testament to the power of seismic engineering, they also represent a gamble that the Taiwanese authorities have taken. Although shaking is very common in the aggregate in Taiwan, the breaking of a fault like what happened in 1999 has seldom caused disruption, damage or loss of life. As a result, Taiwanese building codes prioritised protections against shaking but ignored the risk of faulting as a significant problem. Taiwanese buildings were built so strong that even though their foundations were rent asunder, they were simply picked up and moved out into the streets without collapsing. But because the buildings were not designed to resist the breaking of a fault, they were a loss from a financial point of view, having slid off the owner's property and onto the streets.

Infrequent and ignored

The Chichi earthquake, the second largest in Taiwan's recorded history, happened just twenty years ago. Mount Merapi's latest eruption was even more recent, but is already rapidly receding from our collective social consciousness as we worry instead about election cycles or economic busts and booms on nearer horizons.

When it comes to hazards that are even more infrequent—hundreds of years or a millennium or more in the offing—we have little desire to even consider the risks, much less take action. However, some of these low-likelihood risks pose such a serious threat to our existence that we cannot afford to ignore them or kick the can down the road. The first step, then, is to use science to understand what the risks are, and collect the data with which we can begin to have a conversation about the level of risk we are willing to accept.

And the data shows that the risks are real. Ice core records tell us that every 300 years or so, there is a volcanic eruption so large that it can have worldwide climatic effects, resulting in food security challenges with the potential fallout of mass starvation and the outbreak of epidemics.

Despite the near certainty of such a big eruption within the next few centuries, we still do not know which of the hundreds of volcanoes in Indonesia are potential culprits, because there has not been a systematic programme to see which ones have large magma chambers beneath them. The reason we still do not know much about the many volcanoes in Indonesia is that the problem is largely regarded as an academic one. However, it is an academic problem that within a matter of a year or less could become a very real problem for much of the world.

On the other hand, there are some infrequent hazards that are so cataclysmic that we probably should not worry about them. About

100,000 years ago, a big chunk of the main island of Hawaii slid into the ocean, rapidly displacing water and causing a trans-Pacific tsunami of epic proportions. Such devastating events have occurred only about three or four times in the couple of hundred thousand years since we evolved to be Homo sapiens. Not only is the likelihood of such a mega-tsunami within the span of human civilisation low, there is also not much that we can do to stop 300-metre-tall waves along the Hawaiian coasts, so worrying about such a possibility is pretty much pointless.

Stretching our horizons

Nonetheless, in between these civilisation-destroying hazards and more mundane everyday risks lie a whole range of scenarios that we can prepare ourselves and future generations for. With the help of science, we are now able not only to map earthquake faults, but also to make calculations about how fast the faults are slipping and how large the earthquakes they produce would be. This ability to make predictions based on an understanding of the past stretches our horizons and helps us to see beyond the limits of our relatively short historical memories.

As a graduate student, I traced the history of the San Andreas fault northeast of Los Angeles back about 1,500 years, discovering previous earthquake-generating fault ruptures in about 800, 1100, 1200 and 1550, prior to the only known historical earthquake in 1857. As we learned more and more about these ancient earthquake ruptures, my colleagues were able to build quantitative models that forecasted how the next big earthquake generated by rupture of the San Andreas fault would spread through Southern California. Based on these models, our communities have been able to evaluate how resilient our police forces, telecommunication lines and public utilities are—all crucial to adequate preparation for the next great earthquake. Without doing the past 40 years of science, we would not have been able to get to this point of understanding the beast lurking beneath us and quantitatively knowing what our susceptibilities are.

So, when the risks are known, we can then take precautions to handle them. The Trans-Alaska Pipeline System, for example, was built to withstand fault ruptures that might happen along the 1,200-kilometre-long oil pipeline. The environmental risk of hot crude oil spilling out of the pipe and flowing into the pristine rivers and landscape of Alaska was deemed to be so great that the American government refused to allow the pipeline to be built unless the engineers properly accounted for the possibility of earthquakes. So at the crossing of the great Denali fault, the pipeline was built in a snake-like pattern on top of Teflon tracks, allowing it to slide and stretch during a fault rupture. When 5 meters of slip occurred during the big 2002 rupture, the pipeline did indeed slide and stretch, and no leakage occurred.

However, the 2002 rupture was a fortunate near miss. Knowing that the last few earthquakes along the same fault had caused a rupture of about 10 metres, my boss at the time made the decision to design the sliders to accommodate up to 8 metres of movement. As a young graduate student, I thought that this was taking too much risk and that we really should have designed the system to withstand ruptures of 11 metres. When a fault rupture did finally did occur in 2002, however, the pipe slid till a mere metre away from the edge of the track, narrowly averting disaster. In this case, the older, wiser person made a gamble and won, demonstrating what can be done when science is applied to ensure safety.

Using the science to finish the race

To move our culture to plan for and promote resilience to infrequent geohazards, we need to keep our science in the public eye. Research helps the known unknowns become knowns. Just imagine facing the future of major climate change without science. What would have happened if Charles Keeling had never begun his curiosity-driven experiment to measure carbon dioxide levels on Mauna Loa? We wouldn't know why climate change is happening, nor would we be able to make predictions about what is likely to happen. Continuing to fund science is thus important if our civilisation is going to have a soft landing when it comes to all of these low-likelihood but high-impact events.

Though funding and conducting good science is necessary, it is not sufficient to effect change. In a research project using corals as seismographs, we studied the occurrence of earthquakes along the west coast of Sumatra over the last 800 years. Starting with a big earthquake in 1350, there have been clusters of earthquakes followed by long periods of quiescence. Since the most recent earthquake happened in 2007 after a 170-year break, this pattern suggests that the 2007 earthquake is just the first of a cluster than we can expect to be completed within the next couple of decades.

What has this research on the Sumatran reefs led to? For one, we are now aware that a large earthquake and accompanying tsunami is likely within the next quarter century. Calculations predict that about half the city of Padang will be inundated by the coming tsunami. That suggests that about half the population of about 800,000 are at risk. However, very little has been done to change building codes or to widen streets so that people can be evacuated quickly. This is an example of a lot of good science not getting adequate traction in the world of potential consequences. If we were to think of getting people to act as finishing a marathon, then I would say we are only about halfway toward the finish line.

One way that we are trying to finish the marathon at the Earth Observatory of Singapore is through projects like the one we have in Phuket. Together with the government, local stakeholders and businesses there, we have set up a three-pillared programme that includes a smart disaster-information system, a non-profit enterprise using tourism to raise funds for resilience and finally a disaster resilience fund that would enable locals to build back better. This integrated approach will allow the community to ensure that response plans are effective and that there are adequate resources to respond after a disaster strikes.

The point is to make science part of our cultures, building in a desire to respond to risks in a rational way. Given how we have evolved to prioritise short-term risks over low-likelihood but high-impact risks that are harder to comprehend, this is a serious challenge. But just as we have learnt to cope with frequent hazards as a society, I believe that we can learn to look beyond the immediate and find the courage to deal with our low-likelihood, high-impact hurdles head on.

Kerry Sieh is Director and AXA-Nanyang Chair in Natural Hazards of the Earth Observatory of Singapore at Nanyang Technological University, Singapore. He studies the geological record to understand the geometries of active faults, the earthquakes they generate and the crustal deformation their movements produce. His early work studying geological layers and landforms along the San Andreas fault led to the discovery of how often and how regularly it produces large earthquakes in southern California. More recently, his investigation into Taiwan's multitude of active faults showed how their earthquakes are continuously creating the mountainous island.

NINE MEALS FROM ANARCHY

The fragility of food systems in
a globalised world

By Tim G. Benton

S ince the end of World War II the global food trade has developed to deliver ever more food at an ever cheaper price. But our close interconnectedness masks a fragility that few are aware of.

Our food system is so much more than a conveyor belt of calories; it underpins how our societies work. Food production profoundly influences our environment, our culture and our family life. Food even functions as a bank in developing societies, where families store their entire wealth in a single cow.

By the same token, risks to the food system cut right across society. Malnutrition in the form of over-consumption is a massive health issue, especially in developed societies, while malnutrition in the form of under-nutrition makes both life and social order precarious, especially in developing countries. Food problems are big problems.

When the food system is disrupted, things can go very wrong. Curtailing food availability is the fastest way to create social breakdown. The saying by American journalist Alfred Henry Lewis that "there are only nine meals between mankind and anarchy" is more than just an adage. Only recently, rapid food price inflation fuelled the disorder which sparked the Arab Spring and the ensuing migration crisis, which in turn has disrupted the European Union. This is but one example of the hyper-connectivity and fragility of our food system that we are only now beginning to understand.

The growth and extent of the globalised food system

The food system encompasses the totality of production, how it affects our society and environment, the way produce travels through the supply chain and, in the rich world at least, ends up in a shop. Our choices in those shops affect our nutrition, health and wellbeing. Along the way, waste is created. It is a whole system of loops and feedbacks of demand and supply.

Importantly, it is also a spatial system. The location and movement of food around the world has changed profoundly since World War II. Singapore is at the extreme end, with 90 percent of its food coming from 30 other countries, but almost all nations rely—to a greater or lesser extent—on global trade to feed local people.

It is no accident that every country has come to rely on food from elsewhere. The growth of global food trade was a deliberate policy driven by the US and the UK's desire to open up markets for economic growth, and to create more consumption by making food cheaper. But we have now become so accustomed to the system that it seems based on imperatives rather than choice. It is part of the larger mindset that we live on a planet with finite resources but expect infinite economic growth. As economic growth has increased exponentially, with it has grown the global movement of food, as well as—and here is the first big risk—our reliance on food transport.

Choke points, concentration and homogenisation

The complex web of food traded between countries resembles the financial banking network just before the 2008 crash: high connectivity, and a few large players exporting to many small ones. So, if something happened to a large player, it would knock on through the network, creating a ripple effect.

In addition to the fragile network of food trade, if you traced all of the food trade routes on a globe, you would see a complex web entwining the whole world, but with the threads pulled tightly together into several very powerful nodes. The Suez Canal is one; Singapore another. These nodes, or choke points, are crucial for global transport, and if something goes wrong, the transport of food would fail. A fragile food trade network is overlain by a fragile food transport network.

These choke points funnel not just raw materials such as meat and grain, but also fertiliser for crops, and processed foods. Half of China's soy passes through Singapore. A quarter of all cereal travels through the Suez Canal. Blocking the Suez Canal would have the effect of halving supply, because it would take each boat twice as long to make a detour around Africa.

A world map of where this food is travelling from would show you that the world's calories come from farming in a small number of 'breadbasket' regions. This concentration in space is thanks to another economic sacred cow: comparative advantage. Countries produce and export what they are best at producing, and buy the rest from other countries. However, the few 'winners' crowd out everyone else, leaving a few production areas where farming is very large scale and intensive.

Each of these breadbasket areas specialises in growing what they can grow well, and what they can grow is a small handful of commodity crops. Today, over three quarters of the world's calories come from just eight crops: wheat, rice, maize, sugar, barley, soy, palm and potato.

The concentration of crops and calories is driving another significant risk to our society: the global homogenisation of diets. Since the 1960s we have moved from a globally diverse diet to eating more or less the same thing: a similar range of calorically dense but nutritionally poor commodity crops. This homogenisation creates pressures on our health systems that are difficult to overstate. There are now more obese people in the world than underweight. The 'hunger challenge' is being replaced by the 'obesity challenge'. In total, under 50 percent of the human population is now what is considered a healthy weight.

Our food system is looking increasingly fragile: it concentrates risk in space, from the small number of breadbaskets, to growing a small number of crops, to funnelling transport through a small number of choke points. It increasingly undermines, not supports, good health. And, as demand for food grows, its resilience decreases as it gets more specialised, efficient and concentrated.

At the same time, demand for food is growing. The world's population is likely to expand by a projected 50 percent by the century's end. As well as there being more mouths to feed, the demand for calories will rise along with incomes. Already, an average person demands 3,000 calories a day. Supplying these calories requires an additional 6,000 calories to be fed to livestock. Will our system cope?

Growing threats to the food system

We tend to believe that international food trade is a good thing, and generally it is. Food is cheaper and famine is rarer. But we are only just beginning to recognise the serious risks in our food system, which exist at three levels.

The first risk is one of reliance. Almost every country relies on other countries for its food. So ingrained in our thoughts is this reliance that we do not stop to ask what will happen when it is disrupted. Yet our food supply could be disrupted by small events anywhere else in the world. That is the second risk—the super-interconnectivity of our food system.

The third, the risk of volatility, is the very reason we are finally recognising the risks in the system. In the past decade we have seen at least two big spikes in food price and availability, following a long period of relative stability.

Both spikes were factors in outbreaks of civil unrest. The 2007–2008 price spike helped drive people into the North African streets, sparking the Arab Spring. The 2010–2011 spike created food riots around the world. Some evidence suggests that a long-running drought in rural Syria contributed to migration to the cities, creating social conditions that contributed to civil breakdown in 2011. Hence, the volatility of food prices and the interaction with climate change are absolutely connected to global order.

Several other factors are cumulating into dark clouds that hang over our food system. One is soil degradation. The Green Revolution rapidly increased agricultural yields over the last 40 years, at great expense to soil health. Twenty-five percent of the world's soils are degraded enough to undermine yields, and analysts predict that some areas of intense agriculture only have 50 to 60 harvests left. When fields failed in the past we could simply move on. But studies show little spare capacity in the globe for new agriculture. If we lose a significant breadbasket, it is hard to imagine how we would replace it.

The biggest risk: climate change

Exacerbating the various risks to our food system is the biggest risk of all: climate change. The Paris Agreement commitments, at the moment, put us on course for an expected 3.3-degree Celsius change by 2100. As an analogy, the UK with a climate about 4 degrees warmer is like a whole other country: Spain. Adjusting to this temperature rise would require huge changes in the way we live and farm.

While these changes will no doubt alter our ways, we could gradually adapt. But in the coming decades, a perhaps greater risk lurks which could tip our food system over the edge—the risk of sudden and stepwise climate changes.

The Atlantic meridional overturning circulation (AMOC) is a global oceanic conveyor of heat from the tropics to Northwest Europe, delivering 6-8 degrees of warmth. Scientific modelling suggests that it is as likely as not to turn off over the next two centuries if we do not tackle climate change. There is a small but significant risk that it will cease this century.

If it does, European temperatures would plummet, leading to deeper winters, more storms and shorter growing seasons. Other areas would be affected—the Indian subcontinent might lose its monsoons, and the productive Cerrado in South America would dry out. Food production would suffer enormously on a global scale. And this change could happen over a mere decade.

In addition to changes in climate, our stretched food system is vulnerable to one-off shocks from extreme weather, which is getting both more extreme and more frequent. This vulnerability is not least due to the interconnectivity of the food system. Giant storms in one breadbasket area affect productivity and prices the world over.

In 2017, there were an unprecedented number of Category 5 hurricanes in the Atlantic; in 2015 two of the world's strongest ever storm systems occurred in different hemispheres within a few days of each other. In 2012, maize and soy were hit by drought in the US Midwest. There was a drought in Eastern Europe, yields were hit in Northwest Europe by prolonged rainfall and a very wet summer, and even rice production was impacted by changing rainfall in the Indian subcontinent. Had 2012 been even just a little worse, there would have been the potential for multiple breadbasket failure and significant loss of global production.

Reconfiguring a more resilient system

We are beginning to recognise problems in our food system, but we will likely remain complacent because we are not used to seeing how these problems interact. We need to develop frameworks that are systemic and holistic. We need to break down the traditional silo thinking, where an agricultural person looks at yields, an energy person at oil prices, a health person at diseases, a climate person at weather, and an economist at prices and trade. The functioning of the system, as well as understanding how risks and shocks affect and propagate through it, requires all views combined.

In 2008, a production blip in drought-stricken Australia, a minor contributor to global stocks, caused global prices to balloon. The world's food stocks were already low thanks to high oil prices incentivising biofuel crops over food crops. Exacerbating the problem, governments intervened to protect their people's food security by implementing export bans, which further intensified the demand-supply imbalance. The 2008 experience shows that our food system, which evolved during global stability, is reaching its limits and is liable to break when we hit turbulence.

Governments may have a false sense of security that the market will smooth out food shocks. This sense fails to account for how unprecedented shocks, or tipping points—whether they be the switching off of the AMOC or dustbowl events caused by changing climates and degraded soils in breadbasket areas—might impact global productivity and market responses. Even if some dangers have a low probability over the next decades, they could happen at any time.

The first step in building a more resilient global food system is to recognise the risks and prepare for plausible, if unprecedented, events. Our governments, notoriously plagued by present-termism, need to understand how to respond appropriately rather than lurch into poorly thought-out responses such as food export bans in a crisis.

Research can help us build better understanding and develop red flags. Better modelling of the food production chain will tell us how production shocks translate into short run price impacts. It would also help to identify critical geographical pinch points in international trade, and how to address their vulnerability, such as through investment in infrastructure.

Governments could bolster national resilience to market shocks by, for example, increasing self-sufficiency, building food stocks or diversifying food sources. They should also build international cooperation, for example through rules to limit unilateral food export controls. We should also explore coordinated risk management, with contingency plans, early warning systems and agreed response protocols, as well as coordinated management of emergency and strategic reserves.

Fundamentally, we need to reconfigure our food system so that it delivers nutrition for healthy diets, grown in a sustainable way, from a system resilient to shocks. In the past, we have focused on efficient agriculture and cheap food. But this has created an inefficient food system which is increasingly fragile.

Currently, a third of the world's calories goes to feed livestock, and a third is wasted. Further, about a third of the world overeats by about one fifth each day. If you add these loss factors together, only 40 to 50 percent of the world's food is being used efficiently for healthy human nutrition.

Emissions from the agri-food system are the equivalent of emissions from lighting, cars, travel, washing machines, heating and cooling combined. Changing our diets is a more powerful way of decarbonising our economy than targeting lights or cars. If we halved the amount of meat produced in the world, it would be the equivalent of taking every internal combustion engine off the road.

If we ate less meat, wasted less food and ate less in general, we would make the food system more efficient. This would require less land and water, and would create less emissions and thus slow climate change. We would free up land, so we could farm less intensively. We would enormously reduce the pressure on the food system and make it less fragile to shocks; indeed, the shocks would be smaller as climate change is reduced. Eating less and better would also improve health and well-being for many.

Professor Tim G. Benton is a Distinguished Visiting Fellow at Chatham House and Dean for Strategic Research Initiatives at the University of Leeds. From 2011–2016, he was a champion of the UK's Global Food Security programme, which undertakes systemic analysis and horizon scanning aimed at providing sufficient, sustainable and nutritious diets for all. He has published over 150 academic papers, mostly on the core themes of agriculture's environmental impact, and, more generally, on how systems respond to environmental change.

GOVERNANCE IN A TIME OF DISRUPTION

Why the world needs a new consensus

By Seán Cleary

We face an array of challenges across multiple fields in a highly connected society, and there is an unprecedented increase in the scale of the knowledge we need to address them. It's thus no surprise that decision makers are overwhelmed.

If individual scientists struggle to solve specific challenges in their fields of expertise, how then do we expect presidents and prime ministers—politicians without specialised knowledge in many areas they must address—to choose the right trade-offs in these many interlocking arenas?

The seventeen Sustainable Development Goals (SDGs) agreed upon by the General Assembly of the United Nations as global targets for 2030 provide an indication of the range of challenges we face. Reaching consensus on these goals took a three-and-a-half-year multinational effort, a remarkable achievement that we should celebrate. But the complexity of the goals, and the indicators associated with them, is enormous. It is clear that no individual fully grasps what their delivery across the 193 countries in the United Nations would entail.

The other side of the coin in a complex, highly connected world is what are called global catastrophic risks, events or processes that would lead to the deaths of a tenth of the world's population, or have a comparable impact. The persistence of such risks also derives from the inability of policy makers to address problems of great magnitude timeously and intelligently.

In thinking about how to govern in the face of these challenges, however, we should look to three principles that have underpinned great civilisations, and that have been embodied in all faith traditions, over millennia.

The first principle is the need for a measure of personal freedom within society. This is necessary to enable innovation, because it is impossible for a few people at the top to solve all problems in a complex modern society.

The second principle is that this freedom must be balanced by a sense of obligation to the community, for without this, there can be no society.

The third principle is that each society must respect the ecosystem on which it depends for its survival. The pyramids in the Egyptian desert and the temples in the jungles of Cambodia are monuments to civilisations which at the peak of their success exceeded the constraints of their ecosystems. If we fail to respect our ecosystem, we are at risk of following their fates on a much larger scale.

Economic and demographic pressures in the global ecosystem

Applying these three principles consistently is increasingly difficult because of our unprecedented connectivity. Modern societies and individuals operate on multiple levels—globally, regionally, nationally and locally. We may identify socially with some people from our own community, while identifying ideologically with others, in different parts of the world. These multiple identities make balancing personal freedom, communal responsibility and respect for the ecosystem far more challenging. Indeed, we are not doing particularly well.

Economically, global growth has stagnated since the global financial crisis, except in emerging and developing markets, notably in Asia. The International Monetary Fund (IMF) projects that 2017 will see advanced economies growing by 1.6 percent and the global economy by 3.1 percent. Interestingly, in the past two to three years, the increasing diversity of economic outcomes has made global averages less meaningful. It no longer makes sense, for example, to talk about 'BRIC' growth—the economic performance of Brazil, Russia, India and China has diverged sharply. Developing policies for larger landscapes is harder in this context.

After the 2008 financial crisis, we escaped what could have been another Great Depression by pumping trillions of dollars into the system to keep financial institutions from collapsing. Since then, central banks have used unconventional monetary policies—very low, or negative, interest rates, and large asset purchases by central banks—to stimulate growth. US$3.7 trillion of sovereign bonds presently offer negative interest rates. This policy of monetary stimulus is unsustainable socially and economically. The IMF recommends that monetary stimulus should be complemented by structural reforms, with coordination across all leading economies. We need a better governance model that works for society at large.

Population constraints are also intensifying. Projections suggest that by 2050 cities will teem with 2.5 billion more people, with Asia and Africa experiencing 90 percent of this increase. We shall have to accommodate rising levels of economic engagement as younger populations in the Middle East and Africa join older people seeking to stay in the workforce due to rising life expectancy. We don't know how to accommodate so many generations in the workplace simultaneously, and don't have adequate pension and social security systems to enable retirement.

The growing urban population in Asia and Africa will put further pressures on our environment. While the idea of 'smart cities' with zero net emissions is an alluring solution, deploying the required technology will be difficult as most urban growth will be taking place in underprivileged cities like Kinshasa, Kolkata, Lagos, Dhaka and Karachi.

We are also on the cusp of the largest technological transformation since Industrial Revolution, which caused significant social and political chaos, including revolutions in continental Europe and expansion of the vote in the UK. The present transformation will threaten people without remarkable skills, adaptability and an entrepreneurial spirit. As unemployment and social dislocation rise, representative democracy will struggle to deliver what citizens expect: security and opportunities to deploy their talents to good effect.

Inequality and its discontents

Meanwhile, the shift in the global economic centre of gravity from the Atlantic to the Pacific will continue, as will widening inequality of income and wealth within national economies, due to a 30-year shift in favour of financial returns for capital over labour. Blue-collar wages have been flat since the 1960s. In the past decade, efforts to stimulate economies with low interest rates have allowed people with capital to leverage it enormously. Those without capital have fallen further behind, leading to wealth concentration in still smaller strata.

Although global poverty and inequality between nations has fallen, largely as a result of China's growth, the inequality trend within almost every society has been in the opposite direction. Thirty-four million people, or about half a percent of the world's 7.5 billion people, now control US$112.9 trillion, or 45.2 percent, of the world's riches.

Today's inequality is unsustainable. High levels of inequality correlate strongly with high levels of social pathology. Societies with high social inequality have poorer health, education and child welfare; lower social mobility and levels of trust; and more teenage pregnancies, obesity, drug abuse, violence and imprisonment.

Although we urgently need to address economic inequality, we don't have adequate policy tools to do so. National debt and low growth constrains most advanced economies. India, China and some sub-Saharan African states have high growth rates but have not succeeded in distributing the benefits equitably.

Singapore aims to transform its education system to address these challenges, notably the disruptions that will be occasioned by the new technological revolution. But educational institutions find rapid change extremely difficult, and the insights needed to teach new skills, enable lateral flexibility and ensure lifelong learning are in very short supply. Societies will need flexibility, adaptability, social capital and social cohesion to manage the transition and limit the effects of social dislocation.

The power of populists

In the background of our pursuit of progress we must also take stock of the geopolitical landscape, which has been shifting slowly yet powerfully. Russia's President Vladimir Putin is transparent about his ambitions for nuclear superpower status on par with the US, recognition as a great power and regional hegemony. His domestic popularity has risen with his assertiveness, as his views on Russia's place in the world resonate with a large part of the Russian population. Meanwhile, defective governance and development across the Eastern Mediterranean and the Levant have led to turmoil, sectarian strife, civil wars and waves of forced migration.

Representative democracy is also under threat. National democratic government emerged in a different era with different characteristics, and is under stress in today's world. Individuals no longer need political parties to mediate their interests—Twitter, Facebook and Instagram allow them to express their views—and most political parties have been slow to adapt. Even the current holder of the highest political office in the US circumvents formal institutions when he communicates, seeming to prefer Twitter.

Trust in institutions, moreover, has declined sharply, especially among millennials. Uprisings and protests have spread around the globe, from the Arab world to Hong Kong and from Brazil to South Africa and the Philippines, often exacerbated by governmental malfeasance and inefficiency. Meanwhile, algorithms are coming to supplant individual judgements, notably in security and crime prevention. Although humans still (mostly) make the final decisions—as in drone strikes, arrests and prosecutions— it is increasingly hard to analyse the quality of the algorithmic 'intelligence' that underpins these human interventions.

The opacity of government systems is changing the character and expression of the social contract between government and citizen, threatening the legitimacy of governments. Populist revolts by people who feel disempowered should thus come as no surprise. Those who stand up and say what is felt by many tend to gain appreciable support.

Many populists have common characteristics. They manipulate reality symbolically: they create stories that explain, in very simplified and polemical terms, 'how things work', and propose themselves as the saviours of the people exploited by the elites. They identify protagonists and antagonists, and define a clear-cut basis for intervention. In difficult times, these simple, symbolic narratives appeal more than explanations that 'the world is complicated'.

Of course, once populists gain power, the oversimplified narrative often fails in the absence of policy. To enable their survival, populists then draw 'credible' people from important social, economic and political groups into structural alliances and the ranks of government.

Today we see many examples of populism, from Spain and Greece on the left; and in Hungary, Poland, Austria, Germany, the Netherlands, France and parts of Scandinavia on the right. The UK's Brexit was the result of a populist campaign, as was the election of US President Donald Trump. Some activists are transparently anti-democratic, seeking to undermine the liberal democratic system; others are 'nativists' seeking to prioritise nationals above others, especially migrants; while still others manipulate the democratic process to get elected, and then circumvent the checks and balances in advancing their agendas.

Reframing the scale of risk

The interconnectivity between the threats to our present systems of governance is clear. Geopolitical disturbances leading to large-scale migratory flows challenge representative democracy. Increasing returns to capital, falling returns to labour and the impact of new technologies increase the likelihood of jobless growth. Climate change, exacerbated by higher production and consumption due to more rapid urbanisation, will induce still more migration, and increase geopolitical tensions and pressures on national governments.

Perhaps we ought to ask if we have been guilty of hubris: why did we think we could manage such complexity and rapid change on a global scale? At one level, the deep integration of the global economy, the divisions and fractures in global society, and the inadequacy of political instruments to reconcile these almost guarantees failure.

We need to think about these questions so that we can create a future that we want. The economist Dani Rodrik suggests a need to step back from what he calls hyperglobalisation, and the enormous tension it engenders between economic globalisation and national democratic accountability. He suggests that these are mutually irreconcilable.

We need to recognise that we have exceeded our limits, and that 'more of the same' offers no solution. Instead we must ask what matters of the 'commons' we need to regulate at global and regional scales, and what we should do nationally and locally in order to achieve our goal of sustainable progress within planetary boundaries.

Given the certainty of turbulence as we grapple with this challenge, we need to invest in insight and foresight, to give our institutions—public, private and civic—the ability to anticipate rapid and discontinuous change; and in 'organic' resilience, to enable institutions to recover when shocked by events they did not foresee. Enabling societies to survive unforeseen shocks requires both resilience and social cohesion.

In our efforts to build resilient systems of governance, we can look for inspiration to the design and governance of the internet. The internet was engineered to enable military communication networks to survive a nuclear war. This meant designing for deliberate redundancy, distributed architectures and an autonomous character across all operating systems. This approach may be a useful mental model as we address the larger question of how, and at what scale, to try to manage and govern the complex reality of a human system of 7.5 billion people and rising, embedded in the bio-geosphere, a larger and more complex adaptive system, within whose interactions the potential for unintended consequences is enormous.

Seán Cleary is Founder and Executive Vice Chairperson of the FutureWorld Foundation and a Board Member of the Salzburg Global Seminar. He is Chairman of Strategic Concepts (Pty) Ltd, Managing Director of the Centre for Advanced Governance, Strategic Advisor to the World Economic Forum and a faculty member of the Parmenides Foundation. He is the co-author, with Thierry Malleret, of two books on risk: Resilience to Risk *(Human and Rousseau, 2006), and* Global Risks *(Palgrave Macmillan, 2007).*

THREATS SEEN AND UNSEEN

Terrorism and cyber attacks in Singapore

By Shashi Jayakumar

One rarely thinks about serendipity when it comes to understanding, or avoiding, terror threats. When the Irish Republican Army narrowly failed to kill then Prime Minister Margaret Thatcher in 1984, they told the British security services, "We only have to be lucky once, you will have to be lucky always."

Singapore's security services rely on efficiency, intelligence and foresight. But chance has, on occasion, played a small role in their work. The fortuitous discovery of a video tape in a bombed-out house in Kabul (that of key Al-Qaeda lieutenant Mohamed Atef) in late 2001 is a case in point. The tape, later shared with Singapore, shows a Jemaah Islamiyah (JI) cell surveying Yishun mass rapid transit station in Singapore in preparation for an attack. It is a key reason that the Singapore JI cell was interdicted in December 2001.

It is not that we need luck, nor that it will someday run out. But through a combination of factors—planning, determination, or pure happenstance—it seems a matter of time before a determined extremist gets through Singapore's defences, as robust as they are. It is not without reason that Singapore has seen a pronounced shift in the tenor of security discourse in recent years, with officials at every turn keen to emphasise that it is a matter not of if, but when a terrorist attack takes place.

This certainly has something to do with the fact that the terror threat is diversifying. While JI still has a presence in the region, the majority of those radicalised in Singapore are not members of organised terror groups, but home-grown, self-radicalised individuals, many of whom have imbibed the Islamic State of Iraq and Syria (ISIS) creed. Consider those arrested—and the types of individuals arrested—in the ISIS era. They include radicalised Bangladeshi foreign workers (members of the self-styled 'Islamic State of Bangladesh' detained in late 2015 and 2016); auxiliary security personnel deployed at sensitive locations detained in May 2017; and radicalised young men, one of whom, detained in April 2016, planned to assassinate local political figures in Singapore in the event that he was unable to go to Syria. There was also news in 2017 of another first: the detention of radicalised females. In one case, the individual in question supported the use of violence by ISIS, enjoyed the attentions of ISIS fighters and sympathisers online, and made plans to journey to Syria with her young child in tow.

It is perhaps in recognition of the gravity of the situation that Singapore's Ministry of Home Affairs in June 2017 took the unusual step of releasing a terror threat assessment, with the topline observation that Singapore faced its highest level of threat in recent years. The Ministry's analysis also suggested a quickening of the tempo of radicalisation in Singapore. While it used to take around 22 months for people to become radicalised, the timeline has been shortened to nine months in the ISIS era.

Problems in prediction

Worldwide, political leaders and security services do not have a particularly good track record when it comes to prognosticating terror threats; they often get it wrong. Less than a year before the 9/11 attacks, for example, the US intelligence community suggested in the December 2000 issue of its influential *Global Trends* report that "terrorist groups will continue to find ways to attack US military and diplomatic facilities abroad", with almost no mention of the possibility of the big-ticket, high-signature strike within the US itself.

Sometimes the failures stem from a particular mental model in terms of how things should work. Take ISIS' unexpected advance in 2014 for example. The US erred in its judgement as to how Iraq would work after the former's military withdrawal beginning in late 2017, assuming that a peaceful transition would occur and that sectarian differences could be patched over. These errors of assessment gave ISIS a measure of strategic surprise when it rolled into Iraq. And when ISIS did rise, one of former President Barack Obama's early assessments of the group was that it was a 'jayvee' (junior varsity) team.

Evolutions of the terrorist threat in Southeast Asia have been similarly unexpected in some respects. The takeover of Marawi in the Philippines by pro-ISIS elements in May 2017 was foreseen by very few; few saw, too, that the siege and mop-up operations would last until October. Very little was known of those responsible—in the main, the shadowy Maute Group, which had not hitherto figured high up on the list of likely ISIS co-optees. The security services suffered an intelligence failure. The question for the military and intelligence services (not just in the Philippines) must surely be: How many more Mautes? How many more Marawis?

In Singapore, too, surprises will occur, and certainly in the realm of public perception. In September 2017, pro-ISIS news channels circulated a video clip showing one Megat Shahdan Abdul Samad @Abu Uqayl, a Singaporean ISIS fighter, enjoining others to undertake *jihad*. This raises the spectre of more 'unknowns' surfacing. A measure of comfort could, however, be drawn from the fact that it subsequently transpired that Shahdan had indeed been on the radar of Singapore's security services, with the authorities discussing his case behind the scenes with leaders from the Malay/ Muslim community.

This—the human factor—will be key. The Ministry of Home Affairs has noted that in some recent arrests of radicalised individuals in Singapore, the family (or associates) of the arrested individual was aware of his or her plans or intentions, but did not make a report to the authorities. There can be no mistaking the point: it is the kin-group, first and foremost, that is the first line of defence.

Cyber and hybrid threats

Sometimes, the thinking of security planners is influenced by 'deep regret'—essentially, the phenomenon which sees the fear of big losses or catastrophic events (think 9/11) affect thinking and planning. This can mean that seemingly less obvious threats, or slow-burn issues, are discounted. In Singapore, we should not make the mistake of assuming that a terror attack is the only eventuality we must prepare for. The sobering reality is that terrorism is just one of the many security risks we face. Even in Singapore, where the spectre of terrorism is impressed upon us, cyber attacks of various shades have increasingly affected us.

In 2014, a major cyber attack struck Singapore's Ministry of Foreign Affairs. In early 2017, the authorities announced that the Ministry of Defence's system had been breached, with personal details of 850 national servicemen stolen (the hack, however, did not succeed in accessing official secrets). These are by no means isolated incidents, and senior officials from the Cyber Security Agency (CSA) have made known that Singapore's systems are "constantly probed, regularly attacked and from time to time, penetrated".

It appears to have been a source of concern to the authorities that Singaporeans appear to have relatively low levels of cyber awareness and basic cyber hygiene, with CSA surveys showing a disconcerting proportion of individuals not using basic cyber hygiene when it comes to password management and use of antivirus protection. Again, this is the human factor at work. Many of the key hacks in recent history—one thinks of Stuxnet, which seriously set back Iran's nuclear ambitions for a time—relied not on breaking through firewalls, but on human weakness.

Our ability to build cyber resilience is intimately bound up in Singapore's vision of a Smart Nation. Data-sensing networks,

Internet of Things (IoT) devices, technology powered by AI that help us solve issues from caring for the elderly to improving infrastructure, and data analytics will enable governments to know what is happening and anticipate needs.

All this is a vision holding great promise, but we should be mindful of the perils too. CSA's Chief Executive David Koh has acknowledged that Singapore's high level of connectivity comes with a corresponding level of vulnerability. The type of data a Smart Nation holds is a huge trove for cyber criminals and other malicious actors. Consider, too, the fact that daily life in such a future will be one where everyday objects like fridges and televisions will connect in a mesh: the IoT. Again, a good thing, all in; but at present there is a glaring lack of awareness about securitising backend systems in an IoT-enabled future. And besides the general lack of security awareness and consciousness, no one wants to pay for the security either—not the manufacturers, and certainly not the consumers.

These are just some vulnerabilities of the Smart Nation. Singapore has thus far—at least as far as is known—escaped damaging attacks against industrial control systems and supervisory control and data acquisition systems that have become all too common in the cyber age. It may only be a matter of time before these attacks occur, at least partly on account of the fact that we live in an age of insecurity by design and inaction.

One plausible scenario—possibly ongoing even at present—is of regular, persistent testing of vulnerabilities in the state and private sectors, leading to wider attacks. But probing vulnerabilities may not be confined to technical cyber exploits. Adversaries may try to use sophisticated misinformation campaigns to achieve their aims, employing tools such as influence operations and social media manipulation. These are threats that should be taken seriously, especially in a nation like Singapore, which has a superstructure of multiple races, languages and religions. This diversity is a strength, but needs constant vigilance and tending. Societal cohesion, even (or perhaps especially) in a diverse and cosmopolitan nation can potentially be compromised by an adversary seeking to undermine the polity by leveraging on, and exacerbating, various points of difference.

The resilience dividend

It is not just normal conditions in Singapore that have prevailed through the decades, but 'supernormal' conditions: low crime rates, no natural disasters and no communal violence except what was recorded decades ago in history books. It is extremely difficult on the part of the people to envision threat and post-crisis situations, and just as difficult to cultivate toughness and resilience under these conditions.

What has been useful is the evolution of the government's security approach and messaging. This has changed markedly from absolute prevention to acceptance that something will happen and how to deal with the fallout. For the terror threat specifically, a great deal has been done; take for example the nationwide call to arms known as SgSecure (itself in part a revamp of the decade-old Community Engagement Programme) which seeks to galvanise people in day-after scenarios. This includes promoting ground-up responses to ensure cohesion in the delicate fabric of communal relations should a terror cell carry out a mass casualty attack. The 'bounce-backability' of our society is therefore the present focus. This bounce-backability quotient is determined not so much by government as it is by society. Here, more work needs to be done.

Resilience in Singapore has been underpinned by a historical narrative based on vulnerability (the Japanese occupation during the World War II) and fragility (of race and communal relations for example, and the memory of race riots that took place in 1964 and 1969). Part of the dilemma is that the debate is couched now in existential terms: if modern-day Singapore is hit by a terrorist attack, can we survive? Will there still be trust among the various groups? Will we crumble or cohere?

People think of these as existential challenges and consequently respond in existential ways: survival, or non-survival. But these historical lessons may no longer be the best motivation, and thus may not be the ideal way to persuade young people in Singapore to play their part in the Singapore story. We should move away from a vulnerability narrative to a resilience narrative. This would see young individuals taking an active part—in some cases, even the lead role—in relevant initiatives promoting cohesion. Various promising initiatives going on at the ground level involving interfaith discourse might be taken as a case in point.

This is not to say that government evacuates the security space; far from it. The future role of governments will be to ensure normalcy and physical security in a world—indeed an immediate region—that is becoming increasingly insalubrious. What the government will have to do at the backend is to help prepare society and the people within it to live in environments (real and digital) that are all to some degree compromised. The actual environments themselves—be they real-world or cyber—may be landscapes that we have to accept can never be fully safe and secure. That would be an unreachable utopia.

Shashi Jayakumar is Head of the Centre of Excellence for National Security at the S. Rajaratnam School of International Studies, where he is also the Executive Coordinator of Future Issues and Technology. A member of the Singapore Administrative Service from 2002–2017, he served in the Ministries of Defence, Manpower, Information and the Arts, and Community Development, Youth and Sports. From August 2011–July 2014 he was a Senior Visiting Research Fellow at the Lee Kuan Yew School of Public Policy. His interests include local society and politics, extremism, social resilience and aspects of homeland defence.

THINKING
THE UNTHINKABLE

The role of foresight in US national security

By Sheila Ronis

I n December 2005, the US Congress mandated and began funding an effort to put in place a national security system fit for the 21st century. Called the Project on National Security Reform (PNSR), it had a vision working group, which I chaired, to recommend ways to improve US national security.

Our group used foresight tools to test assumptions and project recommendations. My friends at the Pentagon say the really important part of foresight is opening our eyes and minds to things that we ordinarily would not consider—to think the unthinkable.

Such thinking is the ultimate learning and planning tool, but many organisations today still do not use foresight tools because they do not value learning or the knowledge it brings. If leaders do, however, want to learn, foresight tools can help them. First of all, they would need to acknowledge that they don't have all the answers and that there is a need for new knowledge. Government leaders sometimes find that attitude difficult to accept, as did the Pentagon right after World War II, when America believed it knew all the answers—before it lost its first war, in Korea.

The foresight processes I work with evolved at the end of World War II, when the US Congress asked Herman Kahn of the RAND Corporation in California to help sort through the myriad issues surrounding nuclear warfare. He developed a process to force decision makers into thinking the unthinkable: what would really happen if nuclear war became a reality? Kahn's process led to the understanding that nuclear war meant there could be no winners.

Building on a Cold War legacy

As Kahn developed his thinking process, scientists were beginning to view the world differently. They used synthesis as well as analysis, tools to understand how pieces of a system fit together to make it work. Over time, this line of inquiry evolved into systems science. Scientists also began to see that general systems theory applied to all natural systems.

Kahn's legacy was in helping create the crucial lessons for the world to emerge from the Cold War without nuclear warfare. We need to repeat his thinking process continuously, because if national security is at risk and society is unprepared, the outcomes can be catastrophic.

The US needs to develop long-term, whole-of-government (WOG) thinking and planning as its core capability, as have places such as Israel, Finland, Britain and Singapore. More than 25 years ago, I did some work with the US Army War College. When I read the US' national security strategy for the first time, I assumed it was a subset of a larger national strategy. I was wrong. The closest thing the US has to a national security strategy is a document published by the White House, which is neither sufficiently long term nor a true strategy that links ends, ways and means over time. At best, it is a list of aspirational goals. That summer, I realised for the first time that the US does not develop long-term WOG grand strategies.

Our group is building on Kahn's legacy by recommending that complexity thinking anchor the White House. We have determined that the President of the United States (POTUS) needs a place in which he and his administration can use thinking processes and capabilities to develop and test grand strategy and policy decisions in the near-, mid- and long-term.

We call that place the Centre for Complex and Strategic Decisions, which is being prototyped to anticipate potential futures. It does this by ameliorating complex problems and improving policy and strategy making, through the systems-level integration of foresight and strategic leadership models with complexity science and decision making technologies.

To support such integration, the White House needs to be a learning organisation. As our group noted in our July 2010 report to Congress, the US government needs to develop anticipatory governance, structures and processes to remain resilient. Our group's research on this was reinforced and validated by Leon Fuerth of George Washington University.

Breaking down the stovepipes

Unlike the private sector, the US government has rarely, if ever, used management tools such as forecasting, scenario-based planning and risk analyses. These tools enable everyone to navigate the complexities of an interdependent world better, making everyone more resilient which, in turn, would make the US as a nation stronger.

From December 2008, we took a systems approach to national security. We examined systems similar to that of the US government to look at the interdependence and interactions of all their elements. This helped us understand better how they held together and behaved.

We found that the US government needed to create WOG solution sets for complex systems problems, and the only way to do that successfully was to learn about the systems' issues. We also found that it needed to apply systems thinking to improve its decision making and create WOG mechanisms to break down the stovepipes of government so that these could work together effectively.

In proposing an apparatus to serve the needs of the US well into the 21st century, we asked: what is the basis for rethinking the national security system? How will its future success be characterised? If questions such as these were to be artificially or prematurely narrowed, situations might be misread, which could affect the nation negatively. National security is rooted in the integration of national power elements, including economic, diplomatic and military might. When these are integrated correctly, a nation's vitality is assured, and its ability to encourage positive change globally enhanced.

Vitality and viability

Not too long ago, America's national security challenges related to sub-prime mortgages, diseased birds and automobile emissions. Pilot training rosters were not typically the focus of national security; today, it is clear that they might well have been.

But there are tools to help us think about both the threats and opportunities that a country faces. Threats can be assessed and prioritised based on considerations such as urgency, impact and mitigation options. Opportunities can be assessed on considerations of probabilities of success, long-term sustainability and proportionality. With this approach, matters of national security can be considered as any situation, condition or entity that has the potential to enhance or degrade the viability and vitality of the nation. In other words, a national security system is responsible for—and measured by—the viability and vitality of the nation.

Such a system needs to exist within a complex, adaptive and learning organisation that can anticipate, adapt and address most threats and opportunities. Its people would share information and collaborate horizontally, accommodate unanticipated needs and partnerships, ensure agility amid uncertainty, incorporate ad hoc structures and processes and maintain a long-term view.

In such a system, it is tough to separate geopolitical, social, technological and economic phenomena; all these interact as a system of systems. I would argue that, in most instances, it is a complex system of complex systems—and that is a challenge for everyone because there are limits to how much we can learn or know with any precision. I find it troubling that, although scientists may understand these ideas, many of the bureaucracies we serve are not populated with knowledgeable leaders on this subject. They want and expect us to predict and control the real-world complex systems we work in. But the physicist, sociologist, management professor and policy maker in me knows we cannot do so.

The real world of policy making is a complex system, necessitating learning and planning because, although such a system cannot be controlled, you can influence it if you understand it well enough. As the ancient philosopher Sun Tzu said in *The Art Of War*: "If you know your enemy and you know yourself, you need not fear the result of a hundred battles. If you know yourself but not the enemy, for every victory gained, you will suffer a defeat. But if you know neither yourself nor the enemy, you will succumb in every battle."

Today, no one is big or wealthy enough to cover the world in terms of knowledge or capabilities, so if success is to be expected, nations must actively be learning, planning, anticipating and, most importantly, collaborating with other nations. That is a huge lesson for the US.

Stress testing the system

In the next 50 years, everyone will face extraordinary changes at such an accelerated rate that it will be difficult to imagine. Yet, today's world is one in which many in the West are playing chess while those in the East are playing *weiqi* or Go. Their mental models are so vastly different that they do not know each other in the Sun Tzu sense. And yet managing risk in a world of increasing complexity requires an understanding not only of each individual risk, but also of how different risks interact with one another across all system variables.

With such risks in mind, Congress asked our group to create scenarios that would provoke discussion and hopefully lead to more resilient systems. We created nine scenarios and three timeframes—2020, 2040 and 2060—and also developed a questionnaire. Then, with input from the National Academies, we got 1,500 of the best minds in the US in diverse disciplines to respond to it. We hoped to get 20 people to participate in the two-hour-long questionnaire, but were pleasantly surprised when 133 people from a whole spectrum of disciplines responded.

Our group then crunched these experts' responses and wove the results into our scenarios which, to provoke greater learning, were intentionally inconsistent and often bleak. We also got input on the results from faculty who taught the national security curriculum in three major US military colleges.

We then finalised our scenarios and used these to stress test our five major sets of solutions from lots of different angles, asking questions such as: "How well was the system able to anticipate scenario problems?", "How well was the system able to recover and react?", "Are there problems or solutions identified that we haven't addressed?" and, most importantly, "If this future isn't desirable, what choices should we be making today to avoid it?" The result was that our group's major findings significantly improved the performance of national security systems.

Lessons from Singapore

Complementing our research-based perspective was my personal experience in Singapore as a Fulbright Scholar in 2012. Through studying the Strategic Policy Office of the Prime Minister's Office, I learned how the White House's executive capabilities, including those related to national security, could be enhanced further.

Peter Ho, the architect of Singapore's foresight system and process, told me that complexity thinking has four major roles in Singapore's government: first, to challenge conformist thinking by building ties with international think tanks and global thought leaders through conferences and projects; second, to identify emerging risks by creating risk maps and sharing these with decision makers; third, to calibrate strategic systems thinking to develop new policies and capabilities; and finally, to cultivate capabilities, instincts and habits through systems thinking to deal with disruptive shocks. These roles would enhance the executive capabilities within the office of POTUS—if we could adopt them.

The US needs to be far more proactive in using foresight tools to shape a future of increasing liberty, prosperity, justice and peace, because that is a world our future generations deserve. We hope the Center for Complex and Strategic Decisions will help shape a freer, kinder and universally wealthier world, by informing the policy and strategy that emerges from POTUS' executive office—no matter who the occupant of that office may be.

Sheila Ronis is Distinguished Professor of Management at Walsh College, Michigan, and Director of its Center for Complex and Strategic Decisions. She is also President of The University Group, a management consulting firm and think tank, and Chair of the National Defense University Foundation Board of Directors.

THE RISKS AND RESILIENCE OF CITIES

How disasters can catalyse transformative change

By David Lallemant

Hazards are natural, disasters are not. In the field of risk management, we consider that while events such as earthquakes are natural, the resulting disasters are not—they are the result of human decision in terms of the urban systems we build, where we build them and how we build them. Each of these factors into the level of risk associated with a natural event.

Risk on the rise

Risk is the combination of three components: hazard, exposure and vulnerability. When people talk about growing risk, most understand this as increasing natural hazards due to climate change. But because exposure is increasing, and vulnerability is likely as well, they all get compounded together into a global profile of risk that is increasing much faster than we currently account for.

Hazard refers to geophysical events such as earthquakes and floods, which are typically described in probabilistic terms. Changes to weather patterns along with sea-level rise driven by climate change are increasing the frequency of extreme events such as floods and droughts. At the same time, many of our cities are subsiding due to mismanagement of groundwater.

Exposure refers to the people, infrastructure and other assets that are affected by hazards. This is an important concept—if an earthquake happens in the middle of a desert where there is nothing to be exposed to the hazard, then no disaster occurs. But globally, exposure is increasing because the world's population is growing. The problem is compounded in cities, which absorb the vast majority of growing populations—there are more people living in cities today than there were people on the planet in 1970.

This incredible trend in urbanisation is associated with higher exposure to hazards. Cities are statistically more likely to be in hazard-prone areas, such as along coasts or rivers, in valleys (where fault lines tend to be) or near volcanoes (where soil is fertile). Things get even worse if we look at patterns within cities. As cities get constrained in space, each new city dweller is more likely to settle in an area of higher hazard, often on steep slopes or on low-lying reclaimed land.

Finally, vulnerability describes the propensity of the exposed people or assets to be negatively impacted by hazards. We presume that we are getting smarter in terms of how we design and build infrastructure, but the systems that underpin our cities are also becoming more complex, interconnected and optimised—which in many ways makes them more vulnerable. The socio-technical-natural systems we depend on are increasingly complex, giving rise to new vulnerabilities which we do not fully understand.

Riskscapes and building for resilience

All this paints a rather bleak picture. But one of the reasons I remain optimistic is that we also have a unique opportunity. The incredible process of urbanisation is not yet over. Over the next few decades, in order to meet the demands of our growing cities, we will have to build the equivalent of everything that humanity has ever built in the past 6,000 years. How and where we decide to build this infrastructure will lock us into a risk trajectory for the next century. Hence, we have an amazing opportunity to set ourselves on a path towards resilience—if we decide to do so.

To build for resilience, we first need to understand what I call 'riskscapes'—the changing landscapes of risks in a multi-hazard world. Part of my own work has focused on how the dynamic processes of urbanisation change the risk profile of our cities, in terms of their exposure and vulnerability—two aspects over which we have complete control.

Take vulnerability, for instance. I was interested in the way that earthquake vulnerability is influenced by the process of incremental construction—an ad hoc process where you start with a one-storey building, then expand it or add another storey after a few years, and so on as necessary. This is the default process of city-building around the world, as it is the only way to meet the housing needs of the tremendous numbers of people who flock to urban areas.

Yet, no one is looking at what incremental construction means for risk. This was really driven home to me in 2013, following the spontaneous collapse of the Rana Plaza garment factory in Bangladesh, which killed 1,300 people.

Although this tragedy put enormous pressure on the garment industry to improve working conditions in its factories, there was very little discussion about why a building would collapse to begin with. Rana Plaza's building permit was originally for a four-storey building. When it collapsed, however, its central part was eight storeys tall—and from the rebars left standing in it, it looked like there were plans to add even more onto it.

Choosing a better risk trajectory

This is terrifying when you consider that South Asia is also prone to earthquakes; the next big earthquake could be truly catastrophic, killing upwards of a million people. Yet buildings like the Rana Plaza garment factory are everywhere. Even before the 2015 earthquake,

I was particularly worried about Kathmandu in Nepal. Using structural analysis models of numerous potential incremental building states and a stochastic simulation of these increments, together with models of how the city will expand, I was able to simulate the ever increasing trend in Kathmandu's past earthquake risk, along with its likely risk trajectory in the future. With these models, we can start thinking about what needs to be changed.

Such models of future risk trajectories are empowering for cities whose existing risk is already nearly insurmountable, giving them an avenue to focus on preventing future risk rather than addressing existing risk. With some 100,000 vulnerable buildings, there is very little Kathmandu can do about its existing residential building stock. Massive-scale retrofitting programmes are not feasible, nor is displacing large numbers of people to safer sites. The best it can do is to prioritise critical infrastructure such as schools and hospitals—actions that, to the city's credit, are currently underway.

By using models for future risk, decision makers in cities like Kathmandu are empowered to choose a better risk trajectory. They can test out different policies—stricter construction standards or more stringent enforcement mechanisms, for example—and study their impact on future risk.

The many dividends of resilience

To convince cities to invest in resilience, we need better ways of measuring its true benefits. For example, let's say we want to make the case for Singapore to invest in resilience infrastructure for flood mitigation as sea levels rise. Today, we would typically use actuarial or risk modelling to compare the costs of such investments versus the probabilistic reduction in losses due to flooding in Singapore.

But there are other benefits of resilience beyond the 'reduction in losses' that actuarial tools do not measure. For instance, a city that invests in resilience—indeed, that takes pride in its resilience—is going to attract more investments: companies are more likely to set up their headquarters in Singapore if they know that the city is prepared for the effects of climate change. This is an additional benefit which Singapore will reap regardless of 'reduced flood loss'.

This concept holds true at all scales, including at the level of households—if a family feels that their home is safe from floods, they can invest in their children's education and perhaps take on more entrepreneurial risks, because they are secure in their resilience. Hence, investing in resilience leads to benefits even if there are no disasters.

Another type of benefit of resilience investment that we currently do not measure is that of co-benefits. These occur when infrastructural or natural ecosystem investments bring benefits other than just reducing disasters—for example, a retaining pond that helps to control flooding every five to ten years can also double up as a park that can be enjoyed every day. But as things stand, we are not measuring these co-benefits, and therefore neither are we designing to maximise them.

Maximising co-benefits is important also because the idea of investing solely for the benefit of avoiding losses is a hard sell for most politicians. If a city spends heavily on retrofitting its buildings against earthquakes, people are not likely to celebrate the investment when the next big one hits, because there will still be casualties or injuries even if the retrofitting programme succeeded in cutting the death toll ten-fold. Still, it is important for us to keep making the case for resilience to our political leaders, and to keep emphasising its everyday benefits.

Transformative change: beyond 'building back better'

Resilience is often defined as the ability to bounce back after a disaster. 'Building back better'—the idea that when a city rebuilds after a disaster, it must do so in a way that makes it more resilient— is now the gold standard in post-disaster reconstruction. But in my opinion, this form of recovery is not enough. By focusing on the physical 'building' part of 'build back better', it ignores the processes that create vulnerability to begin with.

I advocate for a form of 'reformative recovery', which goes beyond better building standards to address the systemic processes— issues such as access to land and political disempowerment—that create extreme vulnerability.

There is a precedent for the idea that disasters can catalyse transformative change. The Great Lisbon Earthquake of 1755 almost totally destroyed the city. Some scholars have argued that it also led to the creation of the modern nation state. Portugal, understanding that the economic impact of losing Lisbon would be enormous, took responsibility for the city's reconstruction, with contributions from taxpayers across the nation—the first time in history that this had ever happened. The Lisbon earthquake was also the inspiration for Voltaire's *Candide*, which had a profound influence on philosophical thinking and thought in regard to religion, royalty and cruelty. In this way the Lisbon earthquake catalysed global changes.

Thus far, we have been talking about resilience in a purely positive light. But the processes that create and perpetuate extreme vulnerability are themselves very resilient. The focus on building 'positive resilience' perhaps detracts us from focusing on addressing this 'negative resilience'. This is like combating malnutrition by focusing all our resources on developing food supplements and rehydration salts, while ignoring the larger forces that drive world hunger.

Thus, while basic science research remains critically important, we also need to tackle the systemic and structural issues that are at the root of the problem. In this regard, scientists and engineers need to concern ourselves much more deeply with issues of equality, morality and ethics.

David Lallemant is an Assistant Professor at the Asian School of the Environment, and a Principal Investigator at the Earth Observatory of Singapore and the Complexity Institute, both within Nanyang Technological University, Singapore. He is also Co-founder of the Stanford Urban Resilience Initiative and Co-Risk.org, and has been a consultant to the World Bank and the Global Facility for Disaster Reduction and Recovery.

.

PARADOXES AND PARADIGMS IN PARASITOLOGY

Towards a proactive response to emerging infectious diseases

By Daniel Brooks

Climate change is an equal opportunity phenomenon—it doesn't discriminate for or against anybody, and no human belief system is spared its effects. But as a parasitologist and an evolutionary biologist, I'm particularly interested in the effect of climate change on disease.

Over the last 80,000 years, the Earth's climate has gone through fairly large-amplitude fluctuations—except in the past 10,000 years or so, when there was a highly unusual period of climate stability, coinciding with the emergence of human civilisation. This unusual climate stability has impacted the development of our scientific beliefs, leading Enlightenment thinkers like Isaac Newton to believe, mistakenly, that the world operated like clockwork, and that we could control the laws of nature and even cheat them for our own purposes.

Today, of course, we are beginning to see some major changes in climate, leading to what I call 'The Four Horsemen of the Apocalypse': geohazards, food supply, war and disease. They all ride together, but I will focus on disease, specifically what I call the emerging disease crisis. This crisis is not about a small number of viruses killing a few people in tropical countries, but all diseases, known or otherwise, which affect every species that we depend on for our survival and socio-economic development. These emerging infectious diseases

(EID) could come from anywhere—urban and rural, tropical and temperate—and from anyone. The EID crisis is a national security issue for every country because it affects water security, food security, public health security, socio-cultural security and economic security.

From a scientific perspective, something is very wrong today because new diseases are showing up almost daily—and doing so much faster than long-accepted science would lead us to believe. We are also haemorrhaging money and other resources at a time when the EID crisis needs all the resources we can get. One of the reasons we are having this problem is because our medical and veterinary professionals are ruled by the Hippocratic Oath: do no harm. This, in the context of EID, means 'wait until there is a crisis and then try to do something about it', which is neither economically sustainable nor able to make the crisis disappear.

From a paradox to a new paradigm

Why is there an EID crisis in spite of all of our scientific knowledge, experience and expertise? One reason is that scientists are holding on to a paradigm—a scientific belief—that is not correct. That paradigm, called mutual adaptation or reciprocal adaptation, says that parasites are so strongly co-adapted to their host that they should not be able to switch easily to a new host. But if that were so, diseases would not emerge as often as they do.

This conundrum is called the 'parasite paradox', and is something I have personally witnessed in my field studies. For ten years, I ran an inventory of diseases in Costa Rica in parallel with a colleague maintaining a similar set of inventories in the Arctic. Both of us consistently found that under changing environmental conditions, parasites readily jump into new hosts.

We now know, from deep evolutionary genetic studies, that the phenomenon of pathogens switching hosts easily and producing emerging diseases in the process has happened frequently, repeating an evolutionary pattern that goes all the way back to the Triassic Period.

Fortunately, we have a new co-evolutionary paradigm, called the Stockholm Paradigm, after a group at Stockholm University. The Stockholm Paradigm proposes that specialised pathogens do not need new genetic information to jump into new hosts. Instead, pathogens can use their pre-existing genetic variation to establish new co-evolutionary relationships with the target host, opening up space for genetic exploration and colonising new habitats. Once a parasite or pathogen has explored such new space, it might be able to produce new sets of specialist pathogens.

A moving target

The Stockholm Paradigm highlights the fact that the presence of a pathogen in a host is a result of both capacity and opportunity. No matter how specialised an association between a specific pathogen and its localised host may be, there are other possibly susceptible hosts elsewhere that have not yet been exposed to the pathogen, perhaps simply because it is geographically too distant. So the capacity for expanding into new hosts is always there, waiting for something to create new opportunities—something, perhaps, like climate change.

Pathogens jump hosts in bursts, and every study we have been able to do on this shows that such bursts are always associated with episodes of regional or global climate change. So climate change may trigger opportunities for host-jumping simply by allowing things to move around; if a pathogen lives somewhere wet and it becomes dry, it will move away, and vice versa. Such movements bring pathogens into contact with hosts that are susceptible and that have never been exposed to them before.

One of the implications of this knowledge is that we shouldn't rely on co-extinction to wipe out unwanted parasites. The idea of co-extinction is that hosts and parasites are so tightly interrelated that getting rid of the host—say mosquitoes—will help us eliminate the parasite, in this case malaria or any of the other mosquito-borne diseases.

Tapeworms have put paid to the concept of co-extinction. Twenty-five years ago, my friend Eric Holberg documented tapeworms that are 20 million years older than the birds they live in. These worms were originally living in marine dinosaurs that are now extinct; the parasites, however, did not go away when their hosts died out—they simply moved into something new.

With or without us

The good news about this evolutionary dynamic is that it shows how easily evolution can generate diversity following climate change events, including mass extinctions. Palaeontologists have long known that following extinction events, there is not only a renewal of biodiversity, but also an extraordinary evolutionary diversification. As it also turns out, there is a strong correlation between how severe the extinction event was and the speed and extent of the subsequent diversification: the more severe the extinction event, the faster new biodiversity evolves afterwards.

On the flipside, the EID crisis is simply a manifestation of evolutionary systems' ability to cope with major environmental insults. From an evolutionary standpoint, humans are just one species—the biosphere of over 20 million species will survive with or without us. Over 90 percent of the biosphere is made up of microbes that have been around for three billion years; they've seen it all and have a great capacity for regeneration. You'd have to turn the entire planet to cinders to stop evolution.

As a species, human beings are also not at risk, because there are so many of us, and we are everywhere. I know of people who live up the tributaries of the Amazon, who have no idea what the internet is but know how to hunt and fish; they are going to survive. What is really at risk is us: city dwellers dependent on electricity, clean water, modern medicine, airplanes, solar panels and air-conditioning. We live in a bubble of false hope and denial, and the people who feel most protected against the EID crisis may, in fact, be the most vulnerable to it.

There are four reasons why highly technological cities, where most people live today, are unusually vulnerable to emerging disease. First, they produce nothing and so require a constant inflow of goods. Cut off the supply lines, bring the power system down and within 72 hours, city people will be killing each other for food. Second, cities have high population densities, which means that infected people will come into contact with uninfected people more often than in rural areas. The extreme division of labour in cities leads to a high degree of interdependency, such that even if an outbreak does not kill people, it can nonetheless bring essential services grinding to a halt by making people ill.

Third, cities are safe havens for infectious diseases. Last year, a racoon stowaway on a truck from Ontario, Canada sparked a massive outbreak of rabies in the city of Toronto. Had the racoon stayed in the rural countryside, it would have died alone, unlamented but never infecting anyone else. And finally, the real Achilles heel of most modern cities is that their unsustainably high standards of living depend on undereducated and undernourished people who are essentially invisible to public healthcare services.

Three likely scenarios

You might think that the way out of the predicament we find ourselves in would be to stop all the things that we are doing wrong. But we cannot do this because humanity is now living in a heavily constructed niche; undoing our technology-enabled niche would be a massive disaster for us all.

Take, for example, vaccines, which have saved countless lives. The unintended negative consequence of vaccination is that the human population is now made up of a lot of people who are not genetically disease-resistant; they are only technologically resistant. We have created a situation where pathogens and hosts are asked to adapt to our technological niche rather than to each other.

Over the past year, my colleagues and I have been thinking about what would happen when the bubble finally bursts. We have come up with three possible scenarios: two of them hard landings with bleak outcomes, and one softer landing that might happen if we take action now.

The first hard landing would happen if we had a pandemic in a rural, food-producing area. Not only would this directly kill people there, but it would also starve city dwellers of food. The second hard landing would occur if a pandemic wipes out everyone in the city and destroys all our technological infrastructure. Although people in the rural areas survive, they would exist only as non-technological agrarian societies.

A softer landing from these doomsday scenarios might happen if we manage to form a network of cooperating countries that includes producers and consumers—technology and food-producing hubs that are willing to share information and resources with each other for a common good: survival.

Adopting the precautionary principle

The word 'cooperation' can be used in two different senses: an open-ended one loosely committed to finding a solution, or a rallying cry centred on an action plan. To be proactive, we have to replace 'Do no harm' with the precautionary principle, which says that having incomplete knowledge is not a reason to avoid doing the right thing.

It's possible that everything I've just discussed never comes to pass, that there never is any kind of EID crisis. If that is the case, the worst thing that could happen is that you would laugh at me for over-reacting, and I can live with that. The alternative, however, is that my grandchildren get sick and die because you didn't listen to me. Worse still, we may face a situation that humanity never recovers from.

Think back to those temperature fluctuations from 10,000 years ago. If human civilisation could only have emerged during a window of unusual climate stability, we may not get another chance. If we lost all our technological infrastructure, we would not be able to rebuild humanity based on petroleum because the only petroleum left in the ground requires technology to get it out. Solar is not the solution either because we can't make solar panels without petroleum-based technology right now.

We can't medicate, vaccinate and eradicate our way out of the EID crisis, medicating the sick, vaccinating the young and eradicating pieces of biodiversity associated with the transmission of pathogens. We've been doing these things for a long time and they are no longer working. Pathogens have been far more successful at finding us before we find them; we need to reverse that situation. Although we can't stop emerging diseases, we can anticipate them and mitigate their impact, buying time while we work on coming up with real systemic solutions. We have the ability and the impetus—there may not be a second chance.

Daniel Brooks is a Professor Emeritus in the University of Toronto's Department of Ecology and Evolutionary Biology; a Senior Research Fellow at the Harold W. Manter Laboratory of Parasitology, University of Nebraska State Museum; a former Visiting Senior Fellow at the Stellenbosch Institute for Advanced Study in South Africa; a former Visiting Senior Fellow at the Institute for Advanced Study Köszeg in Hungary; and a former Senior Research Fellow at Universidade Federal do Parana's Department of Zoology in Brazil. He is also a Fellow of the Royal Society of Canada.

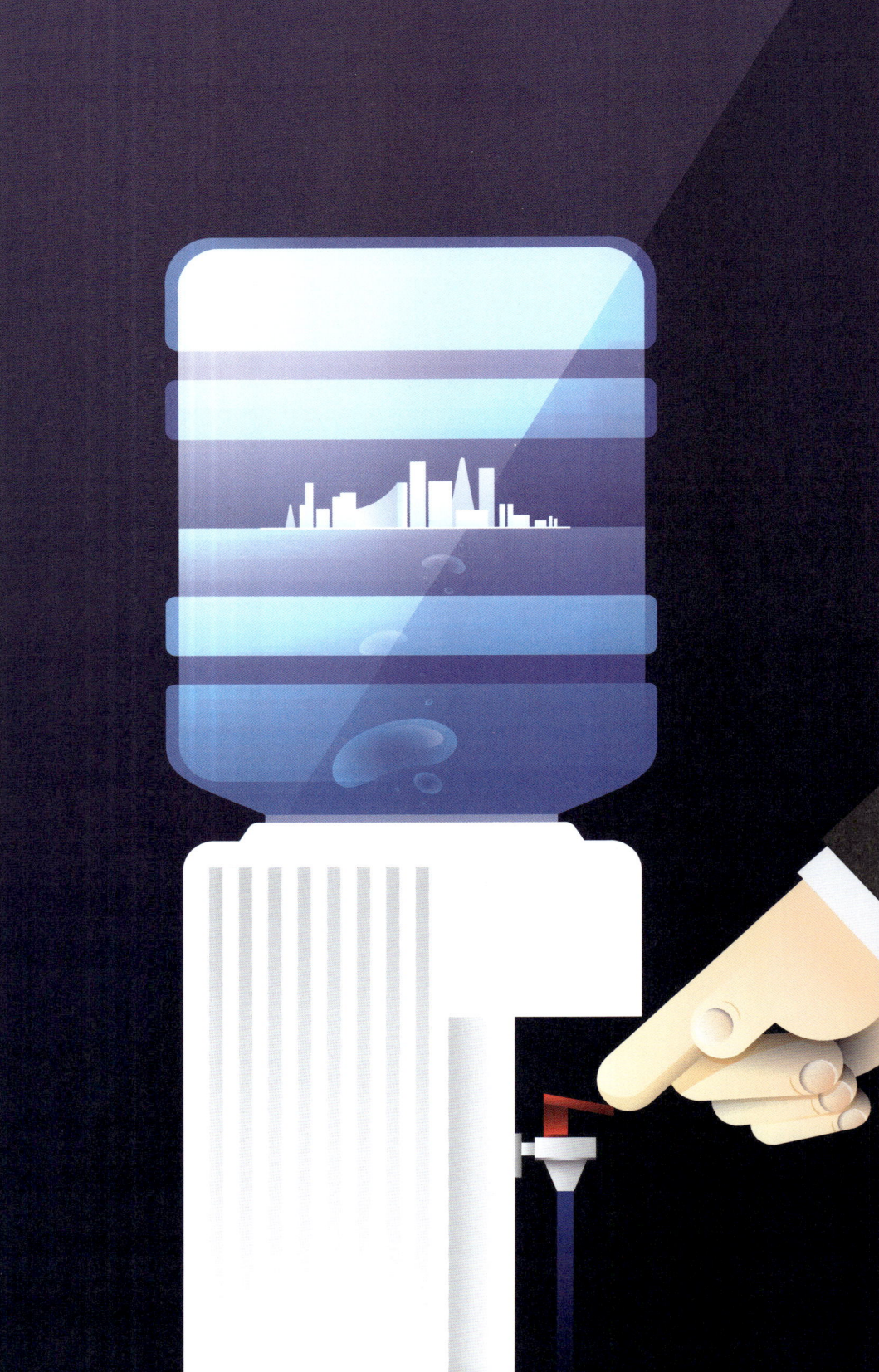

WATER:
THE HUMAN ELEMENT

Putting people at the heart of water policy

By Alexander J.B. Zehnder

Water, with its ability to cleanse and centrality to life, is at the core of major religions and beliefs around the world. However, because issues surrounding water are so emotionally evocative, obvious solutions to the water challenges we face are sometimes not taken. In truth, we know how to provide enough high-quality water for all, and can do so in relatively simple and straightforward ways. But first of all, there must be more individuals and countries to deal squarely with the major challenges to their own and their countries' water supply.

There are two reasons our global water supply is under pressure: first, there has been explosive population growth, so an ever-growing number of people are tapping into a finite volume of fresh water, consuming more as they grow wealthier.

Second, global warming is affecting the global water supply, triggering floods and droughts that destroy crucial water infrastructure, negatively impacting agriculture and natural ecosystems. We could reduce our carbon emissions, but the political will is lacking to do so. The better and most pragmatic way is to adapt to the changing climate, but even that is challenging as the changes are hitting us much faster than before and at much higher cost.

Distinguishing blue from green

To understand water scarcity and why it is a central issue in global change, we need first to understand the quantitative water needs of humans and to distinguish between the two main types of useable water we have. The first is blue water, which we drink and wash with—in short, the liquid water. The second is green water, moisture locked in soil and only accessible to plants. Green water cannot be converted to blue water except at very high energy costs. There is also faeces-contaminated blue water, called black water, and blue water contaminated by households and industry, called grey water. But we disregard black and grey water because they can be treated to become blue water again.

As a rule of thumb, whenever rain hits the ground, two thirds of it becomes green water while the remaining one third is blue water. Except for plants, life on Earth depends exclusively on blue water.

Given the limited supply of water, it's important to know how much water each of us needs to survive. A 1992 study determined that a sufficient amount of water per person per year is between 1,500 to 1,700 cubic metres.

Of that 1,700 cubic metres, you will need three litres a day to drink if you live in a temperate zone and about nine litres if you live in the tropics and work outside the whole day. That translates to between one and three cubic metres of water per person per year, which is a small amount—except that it has to be water of the highest quality.

Though we have technologies for treating contaminated water, spanning high-tech to simple solar disinfection methods, about two billion people still do not have access to clean drinking water. The reasons are manifold, ranging from the inability of leaders to manage larger water infrastructure projects, to the lack of understanding on the part of many of how to safeguard water resources, to the unwillingness to pay for clean water just as we would for any other good.

Eaten, not drunk

Besides drinking water, we need approximately 30 to 50 litres per person per day for cooking and bathing—roughly 20 cubic metres per person per year. Commercial and industrial water use takes that figure up to approximately 250 cubic metres per person per year. An average US individual uses more, up to 370 cubic metres a year, while an African makes do with 25 cubic metres a year, which is close to the minimum volume of 20 cubic metres needed for a person to survive.

Still, an average of 250 cubic metres of blue water per person per year is far less than the lower estimate of 1,500 cubic metres per person per year. Where is the rest of the water going to? In a word: agriculture. It takes 1,000 litres of water to grow the wheat for a kilogram of bread. Meat is even more water intensive. For the same amount of calories, meat requires ten times more water than plant-based food. To produce the necessary food (2,500 kcal per day) for a vegan, roughly 800 cubic metres are consumed; a diet with 20 percent meat raises this number to 1,200 cubic metres. We are basically eating our water, not drinking it.

Our second water challenge is finding enough water to grow food for all—growing plants uses up a lot of water. Although some crops require more water than others, all require primarily green water, except for irrigated crop. In the 1990s, for example, when China struggled to grow enough food for its population, it switched from cultivating waterlogged rice to planting primarily maize, which uses water more efficiently. China did so in response to a study by the World Resources Institute and, in the mid '90s, China's food imports fell to virtually zero.

The real impact of virtual water

Apart from reducing the amount of water needed to produce food, we must be mindful of virtual water, or the amount of water embedded in food that countries import whenever their own water availability for local agriculture falls below roughly 1,500 cubic metres per person per year.

Singapore, for one, still imports close to 800 cubic metres per person per year of virtual water despite its efficient storing of water from rainfall (blue water). Singapore's dependence on such virtual water is at least as much as its dependence on water coming from Malaysia.

It then becomes important to ask: Which countries account for the lack of food around the world? The answer is that there are only five main food over-producing nations: the US, Australia, Argentina, France and Canada. Contrary to what economists have long believed, it is mainly the industrialised countries and not the developing ones that have been and will continue to supply the missing food for the world.

Such a food oligopoly is a considerable power factor. Producer nations can use food export as a geopolitical weapon, and only supply to compliant nations. This weapon has been used in the past and there is no reason why it will not be used again. The main flow of virtual water goes from the five nations to South America, Africa, Russia, India, China and Southeast Asia. The movement of food to these regions makes up a considerable part of the global virtual water trade.

Shifting distribution

At the end of the last century there was a heated debate about whether water should be declared a human right or remain entirely an economic good. It was only in 2010 that the UN General Assembly adopted a resolution recognising access to clean water and sanitation as human rights. The amount of water per person was not defined.

The basic water needs of a human—for drinking, cooking and hygiene—amount to about 20 cubic metres per year. While this amount can be considered a human right, we need to find an economically sound price for non-essential water consumption, be it for filling a swimming pool, watering a lawn or washing a car. A few countries, among them South Africa, have water tariffs which reflect this difference. The water used for basic needs is taxed at a low rate; beyond this, tariffs increase markedly.

Water has no price. Everyone can get a bucket of water from a lake or a river without being charged for it. However, there are costs to getting high-quality water delivered to our homes 24/7. In only a few places around the world are these costs covered by tariffs which really address the cost of production and infrastructure. In most places, water is subsidised to varying degrees, ranging from a little to almost entirely.

Yet, putting a price on water still cannot secure the global water supply needed by the world's growing population. We must better manage what is locally available, but also allow virtual water to flow freely without restrictions of any political nature. Many countries, such as India or China, theoretically have enough water for their populations, but would have to retain every raindrop within their borders. This would mean damming all rivers, preventing all run-off to the sea and so on, which is a highly demanding country-wide management job with expensive infrastructure requirements. Since this is not immediately around the corner for most countries, an optimal adaptation to the severe challenges is the only choice. For those who cannot adapt, solidarity with the global community is the only way of survival.

Climate change will create losers and winners. In parts of the US and Africa, Southern Europe and particularly in India, rainfall will be reduced or will come at the wrong time. Other places like northeast China and Argentina will get more rainfall on fertile soil, allowing higher crop production. Overall, water will be sufficiently available under future climate change scenarios, but the overarching challenge will be managing it right—locally, regionally and globally.

Managing with what we have

The future water challenges are enormous. Most countries will not be able to deal with them alone, be it for the lack of technology, missing management capacity or the absence of will to plan and invest long term. The inability to solve water issues is not restricted to certain parts of the world. A good example is municipal infrastructure for drinking water and waste water. This infrastructure is very expensive, has a lifetime of 80 to 100 years and requires an investment of 1.5 to 2 percent of the initial cost for renewal every year. Only a few cities like Singapore, Amsterdam and Zurich are taking this task seriously; most, like London and Cairo, are underinvesting and largely disregarding the need for renewal.

China, for example, has to invest about US$100 billion in the next ten years to build the necessary wastewater treatment plants. The European Union requires over US$200 billion just to fulfil its own regulations, and the US requires an additional US$30 billion per year just to keep its water infrastructure running.

Water is highly emotional, and this complicates solutions for water issues. Without the trust of the people concerned, water-related changes will not be possible. Historically, kings and leaders solved water challenges in the name of their gods, leaving the calming of emotions to priests. It is the combination of excellent water management and good governance that led to advanced civilisations such as those in Egypt, Mesopotamia, the Indus Valley and Meso-America. Failure to attain this critical combination resulted in collapses. Mr Lee Kuan Yew understood this well, and made the efficient management of water Singapore's top priority.

Solving water scarcity requires high governmental efficiency, paired with the utmost transparency and full stakeholder involvement. When countries fail in one of these critical factors, water issues arise even if water is sufficient. On a global scale, so-called economic water scarcity coincides fairly well with low government effectiveness. A fair future for all is built on good water governance and management. Inclusiveness or the human element is key.

Alexander J.B. Zehnder is Visiting Professor and Member of the Board of Trustees of Nanyang Technological University, Singapore, as well as Professor Emeritus of ETH Zurich. He is also Founder and Director of Triple Z Ltd.

SPRAWL AT RISK

Lessons from Angkor, Tikal and Anuradhapura

By Roland Fletcher

More than half the world's population now lives in cities, or to be more precise, in sprawling megalopoli like the East Coast megalopolis of the United States or the Pearl River Delta conurbation that includes the cities of Hong Kong, Shenzhen and Guangzhou. Although these massive human settlements might appear to be an entirely modern invention, they do in fact have historical predecessors, examples that have serious implications for our collective future.

The cities of Angkor, Tikal and Anuradhapura—the epitome of the Khmer, Mayan and Sinhalese civilisations respectively—were astonishingly rich and powerful, not unlike the megalopoli of our time. And yet, for all their sophistication and wealth, each of these cities met the same fate: a period of decline, and finally, demise; eventually, they were covered by forest.

Disjointed when dispersed

Over the past 15,000 years, human culture has gone through three great transitions in terms of settlement sizes. The first transition occurred after 8000 BCE as the settlements of sedentary agricultural communities first grew to more than one hectare. The next great transition, known as the agrarian urban transition, occurred after 3000 BCE, with compact settlements reaching sizes of larger than one square kilometre. With the Industrial Revolution, in the 19th century the size of our settlements once more jumped 100-fold, going to beyond 100 square kilometres.

Although the size of human settlements has grown over time, it is important to note that the residential density of our settlements can vary widely. When you plot the size of a population against the residential density of the society you will see an enormous range, from the small but extremely compact communities of the !Kung bushmen to the extensive low-density camps of the Australian Aborigines, and from large, compact 19th century London to the immense low-density East Coast megalopolis of the US.

Although Angkor, Tikal and Anuradhapura each came from entirely separate ancestries, what they all had in common was that they were large, low-density cities. The critical thing is that there was absolutely no connection between these three low-density cities, and they were each on a terminal path: when these cities went, they were not replaced by a similar mode of organisation and their pattern of settlement permanently disappeared.

In contrast, compact settlements and their networks are extremely durable and there are successions of continuity through time, for example in China and from Mesopotamia to the Mediterranean to Western Europe and the Americas. I argue that Angkor, Tikal and Anuradhapura demonstrate that the survival of giant, low-density cities, with their vast and immovable infrastructure, is extremely sensitive to climate change. In other words, dispersed systems lead to disjunction, while networks of compact settlements lend themselves to continuity.

Astonishing Angkor

Today, Angkor is popularly epitomised by Angkor Wat, the largest single religious building until the 20th century. From the air, Angkor

Wat appears as a 200-hectare rectangle surrounded by a 200-metre-wide moat. This moat is not simply dug into the ground, but is actually a water system with a barrier built around it and a three-million-cubic-metre artificial platform built inside, a testament to the power and wealth of the Khmer civilisation.

But Angkor Wat is just one temple within 'downtown' Angkor, an area which spanned some 200 square kilometres in the middle of the 1,000 square kilometres of Greater Angkor, the immense capital city conurbation of the Khmer Empire and its 750,000 people. If you had flown over Angkor in the 12th century on a nice sunny day, you would have seen something like 1950s Los Angeles, a settlement with a relatively dense centre spreading out to an enormous dispersed landscape of suburb, with light gleaming off the 4,000 to 5,000 water tanks around which people congregated.

Angkor is strategically located to the north of the Tonle Sap lake, a region which floods 7,000 square kilometres of crop land every year. In the terminology of Southeast Asia, Angkor was a *desakota*, or rural-urban area. Within and all around central Angkor were rice fields worked by thousands of farmers. For one temple, the Ta Prohm, over 60,000 farmers delivered 2.5 million kilograms of rice each year to support a staff of over 12,000 people. From Angkor, the Khmer civilisation controlled the single greatest source of rice on mainland Southeast Asia and also key parts of the mighty Mekong river.

How Angkor was 'broken'

Things took an abrupt turn for the worse, however, in the 14th century, when Angkor was hit by severe climate instability. At that time, global climate was transitioning from the Medieval Warm Period into the Little Ice Age, leading to extreme instability in the Southeast Asian monsoon system. The drop in aggregate temperature caused both mega-monsoons and mega-droughts to occur extremely frequently for more than a century.

LiDAR data that we collected in collaboration with the Authority for the Protection and Management of Angkor and the Region of Siem Reap (APSARA) in 2012 revealed extensive erosion damage to Angkor. The extreme monsoon rainfall was trapped by the city's massive infrastructure and channelled into canals which could not contain it. This damage took out the major water system of Angkor, gouged out the main connector canal running south through the city and washed all that sand southwards, burying the huge downstream canals in the process.

Combining our data with archaeology and dendrochronology data, it is clear that Angkor's water system was disrupted by huge volumes of water that it was not designed to carry. Once the system ruptured, Angkor no longer had the capacity to protect its people from flooding or to ensure water for agriculture. The end result was that Greater Angkor was largely abandoned. In addition, the entire surrounding heartland of the Khmer Empire, in the northern half of modern Cambodia, reverted to forest and villages. People shifted out to small towns on the periphery of the heartland south of the lake and down to the southeast, forming communities that were mercantile-oriented rather than state- or religion-oriented.

That said, you can be sure that the Khmers never lost anything quite as large as Angkor or for that matter, Angkor Wat. The Khmer state maintained its integrity even after the demise of Angkor; it simply moved sideways and down to the southeast to the region around Phnom Penh. By the 16th century, the Khmer rulers were back in Angkor and re-gilding the towers of Angkor Wat, which by then stood in a gradually reforesting landscape.

Cities in the face of climate change

Like Angkor, the end of the Classic Maya low-density city of Tikal also coincided with a period of climate instability in the 9th century. Although the decline of the city has long been assumed to have been caused by drought, it actually falls within a period of extreme temperature change. Ultimately, the collapse of Tikal and other Classic Maya cities led to a gradual movement of the population to small, compact mercantile settlements around the periphery of the old Maya heartland.

This characteristic pattern of moving out of the heartland and into the periphery was also repeated in northern Sri Lanka around Anuradhapura, one of the great low-density medieval Sinhalese capitals of Sri Lanka. Although we still do not fully understand what was happening to the local climate at the time as we do not have a direct proxy for temperature in Sri Lanka, we do have an indication of the temperature changes in mainland India. After the 12th century and following the demise of Pollonaruwa, the partner city of Anuradhapura, small urban centres developed, primarily around the coasts of Sri Lanka, and later up in the hills under the influence of the Portuguese.

If you take the three examples of the great cities of the Khmer, Maya and Sinhalese together, low-density cities appear to be extremely vulnerable to the effects of severe climate change. Climate change directly or indirectly disabled some major component of the massive infrastructure of these cities. The key problem was that once those systems broke, they were extremely difficult to recover.

In the case of the Maya, small states disintegrated into mini-states. The small empire of Anuradhapura, which controlled the entire island of Sri Lanka, was somewhat more enduring, breaking down into numerous small states. Angkor, on the other hand, was the capital of the biggest empire in Southeast Asia. The Khmer rulers and their society adapted and persisted, although they went from ruling an empire to being a regional state. Very large low-density cities are vulnerable, and when they break down, take out their entire heartland region. Though this was surely disastrous for many people, it is important to note, and of consequence for our future, that the affected societies were resilient and enduring.

Looking back to move forward

Extrapolating what we have learnt from our three urban examples to today, you will see a disturbing resonance with the present. Very many of us live in giant, low-density cities. We are heavily dependent on massive infrastructure, and like the Khmer, Maya and Sinhalese civilisations, we have substantially altered our landscape. Further, we are confronted by a period of intense temperature change which is already producing severe climatic instability.

Should our megalopoli go the way of the low-density cities of old, we would see populations migrating out of places like the East Coast megalopolis and the Pearl River Delta, and into small, compact cities far away on the present-day periphery of our major urban regions—out in western and northern China and in central western North America. The effect would be to create a whole new world, one that would be unrecognisable.

The broader point is that the form our cities take is not neutral. In itself, the material form is a 'player' in its own right, affecting our social capacity to adjust and survive. In our world and at the scale at which we construct our environment, the material is an actor without intent that can act against us. What we need in order to cope with climate change are models which systematically incorporate this non-correspondence between materiality and sociality.

The examples of Angkor, Tikal and Anuradhapura are significant because they ended with similar outcomes despite having very different societies, economies and local environments. The Sinhalese state was managing water directly, while the Khmer state does not appear to have done so. Both grew rice. Unlike both groups, the Maya did not have the wheel or big quadrupeds, and grew maize, squash and beans. And yet all three of these once-great cities eventually collapsed and saw the disintegration of their heartlands.

The lesson here is not just that low-density cities are inherently doomed. Rather it tells us what needs our serious attention, though we have never before envisaged so severe a risk. Knowledge of what brought down our predecessors' cities is a gift they bequeathed to us through their tragedy. Can we learn in time what we might do to mitigate our own risk and change our behaviour?

It is surely our job to mitigate disaster and minimise tragedy to whatever degree we are capable. To do so we must pay attention to the past, not treat it as an incidental triviality amidst the daily demands of our urgent present. Though the demise of the previous great, low-density cities of humankind is an ominous story, the endurance of the people and societies that went through that experience tell us there is also much to learn about how to survive. Human beings are extremely resilient, and we will survive, particularly if we heed the lessons of the past.

Roland Fletcher is Professor of Theoretical and World Archaeology at the University of Sydney, and Director of its Angkor Research Programme. He is also a former Visiting Fellow at Durham University's Institute of Advanced Studies.

THINKING THE UNTHINKABLE V2

Why leaders must challenge longstanding assumptions on how to lead during the new disruption

By Nik Gowing

Today, leaders in governments, corporations and institutions face a whole new landscape of disruption and uncertainty. Increasingly, the present and future is a world of new unthinkables and unpalatables. It is a level of uncertainty that few have ever experienced.

Indeed, the world is now moving much faster and in directions which most leaders are not prepared for. In too many surprising respects it is nothing like what their careers had trained them for. Whether in the corporate, government or institutional sectors, leaders are not willing to embrace this reality. In too many ways they are getting derailed by it. This is because their mindset is largely conformist and conditioned by how they got to the top. As many of them have admitted candidly to the Thinking the Unthinkable research project on leadership, it is not appropriate to the level of disruption that is going on.

Leaders are certainly not prepared for it. What we're seeing is an unravelling of much of the glue of institutional frameworks and relationships between countries and corporations in ways which no one has ever experienced.

A main insight from the ongoing Thinking the Unthinkable project is that this is like a new wartime. It is a destabilising war created by a whole raft of disrupted ideas and principles that no one has predicted or expected.

It is not a war of weapons, and it is not just about cyber warfare. It's about the unravelling of what everyone has expected would be stable since World War II and the Bretton Woods agreements of the late 1940s. The stability, reliability and strength of international agreements and treaties are not only being challenged, they are being actively undermined.

And you can date this unravelling to 2014. You can see it when Russian President Vladimir Putin decided, in violation of international treaties, to essentially invade Eastern Ukraine by proxy and seize Crimea. You can see it in the new relationship of the People's Republic of China to other parts of the world, particularly with the islands in the South China Sea. You can see it challenging everything that everyone assumed was guaranteed in terms of stability and predictability. That is what is unthinkable. But it's also unpalatable.

Yet, most leaders in governments, corporations or institutions are still not prepared to understand the enormity of the challenges to everything they have long been prepared for and assumed. After four years of study, the main finding of the Thinking the Unthinkable project is that the conformity that gets you to the top in many ways now disqualifies you from understanding and appreciating the enormity of disruption and its implications, and how to handle it. And I can tell you no one—absolutely no one, even at the highest executive levels—rejects this finding. Indeed they nod their heads and encourage us to keep the research going. They want to know what solutions there might be. They confirm how much they need them!

Leadership under pressure

What we have identified is something which every leader is very concerned about. The words they volunteer to us are 'scared' and 'overwhelmed'. That is what is quite remarkable—that they choose those words.

Additionally, there are new pressures from the next generation. Many don't like what they see. They don't like the way companies are working. They don't like the way governments are working either. And they are pushing back. Even those who have been on MBA courses to get qualified for leadership are saying they don't like what they see. They have second thoughts. Many prefer to embark on start-ups instead of a traditional corporate or government career.

There is an existential challenge for leadership. This is because at the moment, leaders are essentially of a certain generation. They're in their early or late 50s, or maybe in their 60s. Many of them are still clinging on, to put it bluntly, to keep their wealth and their share options, especially in the corporate world.

The situation leadership now finds itself in is a bit like an aircraft which has hit volcanic dust and lost its engines, and is now cruising downwards under the influence of gravity. Things are under control for the moment. But what happens next? Many do not want to accept the enormity of the change that is already here. But more importantly, like the aircraft with no engine power, they face unthinkable change of an even more profound type that is already coming down the track.

And in many ways, you haven't seen anything yet. Artificial intelligence (AI) and algorithms are already decimating the normal assumptions of stability within countries. The strategic challenge now is a threat to social stability. Jobs which many people assume will be there to enable them to enjoy their lives, or at least to have enough to feed their families, simply won't exist. Many jobs in the retail, banking and financial sectors will be replaced by AI and algorithms.

Facing up to the unpalatable

What evidence do we have of the impact that these unthinkables can have on leadership?

Look at the migration crisis which hit Europe in 2015. We were warned about this for at least two years by the United Nations High Commissioner for Refugees, and also the International Organisation for Migration. They both warned European governments that their nations would face a significant migration problem because of the war in Syria and the threatening exodus from north Africa. But most leaders didn't believe the migration crisis would happen. It was an unthinkable. They were not prepared to believe that it would happen on the scale at which it did.

Now why is this important? The migration crisis was not unthinkable: in reality, it was unpalatable. All those governments should have prepared themselves well in advance instead of panicking in the summer of 2015 when vast numbers of people were put on boats and tried to get in through southeast Europe and Italy. And this has now ended up causing the unthinkable of radicalisation of politics in parts of Europe, and new political realignments in countries like Germany, Poland, Hungary and Italy, with the decimation of traditional parties in France.

In Germany, Chancellor Angela Merkel made a serious misjudgement. She barely survived the general election in September 2017. It took her six months to form a new coalition government. This is the price she paid for not being willing to understand the enormity of the migration challenge two years earlier. Accepting one million refugees was a warm humanitarian gesture, but the outcome destabilised politics in Germany. It had the same impact in Poland. It has had the same impact in Hungary, where there is a very nationalist, anti-migrant government.

What's happening is a phenomenon which I wouldn't call populism. It is anti-establishment-ism. People are forcefully pushing back against those who in their eyes have failed. Those who were affected by the migration crisis, those whose jobs and lives were threatened, are blaming their leaders and globalisation. They want things to go back to the way they seemed a few years ago, even if that is unlikely to provide the answers they are looking for.

One of the paradoxes in all of this is that there is now full employment in many countries, even with the threat of AI. But coming down the track, when AI and algorithms remove a lot of jobs, is the prospect of a hollowing out of the middle class. This could then create even greater problems for social stability for many countries, which leaders will be expected to handle.

But are leaders actively thinking about such an unthinkable? And what influence will they have? They have been losing credibility in the eyes of those they serve and those who vote for them. Political leaders have lost control and influence because the public is disillusioned.

Our Thinking the Unthinkable project signalled all of this on 1 June 2016, after we had published our interim assessment warning about the price of conformity in this new era of disruption. We warned that people at the top were disqualified from understanding the enormity of everything that was happening. We predicted the Brexit vote, that Britain would leave the European Union. We predicted that Donald Trump would be nominated, and that he would be elected. Yet, all these kinds of unpalatable, unthinkable risks had simply not been part of the risk calculations of governments and corporations.

A wake-up call

It is a bleak picture. In many ways we have not seen anything yet when it comes to disruption and the resulting global or social impact.

But there could be reasons for optimism. If people accept that conformity is an issue, there are ways now for solutions to be found, or at least, options for solutions can be found. None of them is easy. None of them guarantees success.

But it's about leaders opening their eyes to unthinkables. It's about leaders accepting the inevitability of unpalatables, and having a different relationship with their shareholders, stakeholders and voters. And the most serious challenge, which I've already mentioned, is that the next generation does not like what it sees.

Already, in gatherings like the CEO Initiative from *Fortune* magazine, and to a certain extent the World Economic Forum, you'll hear voices raising real concerns that the future of capitalism is at risk. This is because the public is questioning the right of companies to have a licence to operate. Also, the nature of work and skills has to change because the next generation is saying, "I only want to work for an organisation, company or start-up which is sustainable, which believes in renewables and which believes in a degree of social justice."

These are profound changes. In a letter dated 16 January 2018, Laurence Fink, the Chief Executive of investment management corporation BlackRock, decided to go public highlighting exactly these problems—the need for new purpose for government and new purpose for corporations.

There's also a call for a new set of values, which isn't necessarily about money, but about the value of what you do, and the value to society. That, coming from the biggest financial company in the world, speaks volumes. The same sentiments were echoed by Joe Kaeser, the Chief Executive of Siemens, in an article for the World Economic Forum in that same week.

So the mountains may already be moving, potentially in a positive direction. But there remains an enormous amount of resistance, scepticism and denial. Most in positions of leadership in governments and the corporate world are not prepared to accept these new realities, as was highlighted in the PricewaterhouseCoopers CEO survey published on 21 January 2018.

There are still far too many who believe that what I've described is a blip—a transitory kind of situation, a freak occurrence—and somehow, we're going to return to the way things were. I think that's a flawed planning assumption of governments and of corporations. Some are beginning to wake up to that. But their response is nowhere near enough if you look at the evidence growing daily around us. When you hear President Putin suddenly talking about the new breed of invincible nuclear weapons that Russia is developing, and you see the new power of the People's Republic of China, these are new realities which are completely fragmenting all the assumptions of stability that there have been up to now.

So it's about having open eyes, open ears and open minds. Also, what's needed is a degree of courage, probably a degree of humility and the realisation that in this new wave of unthinkables, things aren't going back to the way they were before. It is now a time of war on ideas and on stability.

The final word is this: you have to be courageous enough to recalibrate leadership skills smartly, at high speed. And you have to build trust. Things are moving very quickly. Leadership needs to be brave and humble enough to realise this so they are not caught out by the inevitable unthinkables that have yet to be thought about. That is when these unthinkables have a massive potential to hit hard and in unexpected ways.

Nik Gowing was a main news presenter for the BBC's international 24-hour news channel BBC World News from 1996 to 2014. In 2016, Gowing co-authored (with Chris Langdon) interim findings of the Thinking the Unthinkable study (www.thinkunthinkable.org). It is based on hundreds of top-level confidential interviews and conversations with corporate and public service leaders, plus hundreds more conversations with the new generation of millennials. In 2014 Gowing was appointed a Visiting Professor at Kings College, London in the School of Social Science and Public Policy. Since 2016 he has been a Visiting Professor at Nanyang Technological University, Singapore, focusing on deepening and widening the Thinking the Unthinkable research.

FROM OWNERS TO STEWARDS

A new way of thinking about finance

By Andrew Sheng

As Deputy Chief Executive of the Hong Kong Monetary Authority from 1993 to 1998, I had a ringside view of the 1997 Asian financial crisis that devastated the region and put an end to the Asian economic miracle. But for all its tragedy, the Asian crisis was ultimately a crisis at the periphery of the world's economy, while its centre—the advanced countries— remained strong. With the 2008 global financial crisis, however, we witnessed a destabilisation of the very centre, with shocks reverberating worldwide.

Although I have thought about financial policy regulation all my professional life as a former central banker, it took the global financial crisis to make me realise that the world had been on a drug all this while—a drug called finance. The global financial crisis shook me to my intellectual core and made me reassess everything I had been taught about how finance works.

Assuming away uncertainty

Where did we go wrong? At least part of the problem can be traced to the end of the Cold War, when many physicists and nuclear scientists moved from working for the government and into finance. They thought that finance behaved like physics, and started making models to manage financial risks. These empirical models eventually became entrenched as financial theory, but in reality they all depended on underlying assumptions that have been blindly accepted.

The fatal flaw of these models was that they assumed that risk is defined as measured volatility, and that volatility itself can be measured, predicted and therefore hedged against. This led to a situation where you could put some numbers into a theoretical model, come up with a figure that defined risk, and use it to guide your decisions. However, these models failed to take into account low-probability, high-impact risks (what Nassim Nicholas Taleb called 'black swans')—the very kind of risks that triggered the global financial crisis.

As far back as the 1950s, English economist Kenneth Boulding warned us not to only talk about risk and forget about uncertainty, which is unmeasurable. But many social scientists failed to heed this warning, collectively deciding that what cannot be easily measured can be ignored, and that what can be ignored does not exist.

The events of the global financial crisis blew all these models apart. The classic example is the Swiss bank UBS' risk management model, which assumed that given the volatility of the markets, the maximum loss would be US$50 million at any point of time. This assumption was proven to be wrong by at least two standard deviations, causing losses to the bank of the order of US$2 billion when the crisis hit.

Despite their flaws, so much has been invested into these models that you still get people trying to sell you the latest value-at-risk models today, built on the faulty assumption of perfect knowledge. Worse still, when you probe deeper to find out what data these models are built on, you realise that the information only goes back to 2000, because that was when most markets started collecting financial market data. What people have been doing is making projections of the future based on very limited information on the past. We trusted these projections, and it cost us all a lot.

Disruptive debt

In the wake of the global financial crisis, we are now in a world that is disrupted not just because of technology, the information revolution and geopolitical issues; we are disrupted because finance is in a massively imbalanced situation. We have forgotten that finance should serve society rather than the other way around, resulting in overconsumption through the over-leveraging of finance or debt.

As a generation, we baby boomers born after World War II have created the greatest wealth, but at the same time also accumulated the greatest debt for our children to pay back. Financial debt is now three to five times larger than the real economy; it has become the tail that wags the dog. This debt is not just from the government but also from state-owned enterprises, private finance and even households. The cumulative debt is huge, and we don't know how to pay for it.

Conventional wisdom says that to deal with debt in a period of stagnation, you should adopt a zero interest rate policy and print more money. The thinking goes that by making money more accessible to borrowers, low interest rates would stimulate the economy by encouraging investment, thereby creating GDP growth and more jobs. However, while low interest rates increase the demand to borrow, they also reduce the supply of savings at the same time by taking away the incentive to save. Low interest rates also create wastage and inefficient investments, causing long-term low productivity and growth. Governments then need to print more money to keep the economy afloat and also its debt sustainable.

But from an incentive point of view, printing money to prevent the present asset bubbles from deflating rewards the greed, excess consumption and leverage that created those bubbles in the first place. Zero interest rates create massive distortions in the system, because it means that commercial banks can make money simply by buying government bonds with no credit risk, whereas lending to the real sector in a climate of deflation could lead to large losses.

But not lending to the real sector creates the exact deflation that the policy makers were trying to avoid.

And that's the situation we are in now—a debt trap. If the US Federal Reserve were to raise interest rates, all the assets we have been paying almost zero interest on might shrink in value. Then, when everybody decides to sell their assets to pay back their debt, prices will crash. This in turn will cause interest rates to rise even further, making the whole situation worse.

A crisis of complexity

As events like Brexit and the election of Donald Trump show, the experts didn't get it. Their models were wrong, and the game has fundamentally changed. In finance, the production game has fundamentally changed as well; everything is now more interconnected and moves much faster and in larger volumes. We are now connected to each other in a web of different networks: urban networks, telecommunications networks, trade networks, supply chain networks and so on. In this complex space, risks affecting one part of a single network can now impact everything else.

Consequently, the difference between the Asian crisis and the global financial crisis was not just one of size, but one of complexity. To deal with complexity, we must try to reduce the multidimensional problem into more manageable parts and study each of them in depth. Although this is absolutely the right approach, people have forgotten the second, equally essential step: we need to take those parts and fit them back together again. We need to see if the whole adds up and is consistent, and not itself imbalanced.

Instead, we have everyone working in narrow, fragmented silos or specialised disciplines, without understanding how each part interacts with other parts, and how this changes the whole system into a more complex, adaptive and dynamic one.

Even the single discipline of economics has become too fragmented, with people specialising in either macro- or microeconomics. As a result, everyone takes a narrow view and makes the false assumption that a small part of the system can represent the whole. The systemic nature of the global financial crisis revealed the total inability of policy makers and financial regulators to see beyond their increasingly irrelevant national boundaries.

By focusing exclusively on the micro and macro, economists have missed out on the meso—the link between the micro and macro— which are the institutions. Particularly in the West, which thinks of the markets as self-ordered, the self-order of bureaucracies tends to be overlooked. Huge bureaucracies like those in China are very often markets in which bureaucrats trade power and information. So those economies are not just vertical markets, but also horizontal ones that are made of layers of networks.

At the same time, economics also missed out on meta-economics—the thinking behind economics. Much of current neoliberal free market economics carries unsubstantiated assumptions or ideology-loaded thinking, such as the idea that free capital flows are self-stabilising. There is increasing evidence, since the Asian and global financial crises of the last two decades, that free capital flows are destabilising to both national and global economies and financial systems.

Modern economic theory is reductionist to the extreme. It states that if you understand demand and supply, the market will work, and the perfect market will lead you to the best solution. This approach completely misses the reality of complexity, and fails to account for the impact of institutions and poor meta-thinking.

New thinking needed

How are we going to cope with this complex, disrupted world? The answer is that we will deal with it exactly as we've dealt with it before—we adapt. But this requires a new way of thinking. We need to abandon the mindset that markets are mechanical and stable, instead recognising that they are dynamic, nonlinear and in continuous disequilibrium. We need to think of ourselves as stewards rather than owners, and move away from a model where finance is the instrument of inequality. We need to think about how finance can begin to serve our community, and not have finance be the master.

Adopting a system-wide view of finance recognises that the global financial crisis was a systemic crisis and requires a systemic solution. It necessitates that we work with different stakeholders, in recognition of the fact that we are all in the same boat.

This generation, of which I am a part, has made an epistemological and philosophical failure by assuming that short-term private greed leads to public good and collective stability. This is a complete misreading of Adam Smith, who many people forget was first and foremost a moral philosopher. It is time for us to allow those who have a stake in the future to take over.

We are now in a process of discovery, and the only way we can discover future truths is to allow the young to share in what we have. If the old theories are wrong, we throw away the theory. If the leadership is wrong, we replace it. The older generation that has made mistakes needs to share both their wisdom and stupidity with the young, and allow them to make their own mistakes.

Andrew Sheng *is a Distinguished Fellow of the Asia Global Institute, University of Hong Kong, and Chief Advisor to the China Banking Regulatory Commission. Sheng was Chairman of the Securities and Futures Commission of Hong Kong from 1998 to 2005, and prior to that, the Deputy Chief Executive responsible for reserves management and external affairs at the Hong Kong Monetary Authority. He has written extensively on international finance and monetary economics, financial regulation and global governance, and sits on several international, regional and national boards and advisory panels.*

FOCUS ON THE FUTURE: A CALL TO ACTION

When we kick hard problems down the road, what we are really doing is passing them on to our youth. They are the ones who will need to tackle neglected risks and suffer the consequences of our ill-preparedness. In what seems like a double betrayal, we are not teaching them nearly enough about their role in diminishing risks and meeting severe disruptions with resilience.

It is our choice whether to leave the next generation roads of optimism, or a mire leading to a bleak horizon. This is the essential message from this book: while strong forces threaten humanity, we have ways of dealing with those threats and avoiding the disasters they foreshadow—we just need to choose to act.

A collective effort to secure the future

Taking action requires the application of science and technology, overcoming emotional and cultural barriers, and the breaking out of silo mentalities. While some solutions lie in the hands of individuals and communities, others rest with local or regional governments; still others require global collaboration.

Individuals can substantially help to reduce carbon emissions and wasted water by bringing what they want in line with what they need for a healthy diet, like reducing the amount of meat they eat.

Communities can act and self-organise locally while mobilising people, money and technologies internationally. For example, to counter the dynamically changing threat of terrorism, they can work in communities of practice where experts from various disciplines come together to share ideas.

Industries can create profitable business cases for possible solutions to the big problems. For example, by working on the demand side as well as on the supply side of the food system, they can contribute by reducing waste and redesigning the system for greater efficiency.

Local governments can bring important issues such as water management to the top of their priority lists. They can also impose new building regulations to increase urban resilience against natural hazards.

International institutions can create awareness and a sense of urgency about the challenges facing humanity. They can also help national governments to take actions that fit their particular interests, scale and circumstances.

Inaction and inertia

If we know what needs to be done, why are we not already doing it?

First, cognitive biases fill individuals' shoes with deadweight. We all have blind spots. We convince ourselves not to see the elephant in the room, and act surprised when it smashes the furniture. We engage in hyperbolic discounting, underestimating the potential of future disaster so we can enjoy the present.

Governments, meanwhile, lower their headlights to cover short-term problems and election cycles, knowing that someone else will be in the driver's seat when it comes time to face the long-term issues. The drug-resistant pathogen with pandemic potential is somewhere further down the road, they hope; thus, the concerted global effort required to tackle it is stuck in neutral.

The nature and dynamics of the changes that governments and international institutions have to deal with—changing demographics, rapid and accelerating technological transformations, growing interconnectivity—are difficult to manage, and at this stage still beyond most governments' capability. At the same time, governments have difficulty relinquishing control to let change happen in an organic, bottom-up fashion.

In short, there is an enormous gulf between what we know can be done technologically, and the incentives and motivations for politicians and decision makers to act.

Singapore's attitude and approaches

As a city-state whose development has followed a survival-against-the-odds narrative, Singapore is almost unique in its intensive analysis of possible futures, and in its no-holds-barred preparation for shocks.

This 'little red dot' provides some outsized examples of the attitudes and approaches needed to cope with sudden disruptions. The 2002 SARS crisis was a case in point—Singapore responded with an aggressive strategy for early detection of infectious diseases and a highly effective response plan to mitigate the huge disruption to lives and the economy. Through prudent planning, the nation has also built strategic reserves as a buffer to disruptions; these reserves helped Singapore get through the 2008 global financial crisis, while still maintaining a significant competitive advantage.

Having weathered a turbulent financial climate, Singapore is also preparing for a time when the skills of today's workforce become obsolete, by creating and adapting infrastructure to deliver continuing education.

While much of Singapore's planning has been government-directed, it is also developing community approaches to handle uncertainty and complexity. A community response plan, for example, is the cornerstone of Singapore's counterterrorism approach. Networks of community leaders and influencers are in place to strengthen understanding between races and religions, and to gain insights into the thinking and reactions of those susceptible to international terrorist messaging.

In short, the Singapore government has not had the luxury of avoiding complexity; in fact, it has made complexity thinking an essential part of its policy making, planning and operations.

Leading the region

Within Southeast Asia, Singapore is uniquely positioned to lead the region in preparing for major disruptions and dealing with their consequences.

Indeed, Singapore has a strong motivation to play this role—the country lies at the centre of a heavily populated region that frequently bears the brunt of earthquakes, cyclones, flooding and tsunamis. While Singapore, by quirk of geography, is largely free from the direct assault of natural disasters, it is extremely vulnerable to their secondary effects, the severity of which will largely be determined by the disaster's impact on directly affected societies, and on how these societies respond.

Thus, the ability to anticipate and quickly mitigate the consequences of major disruptions will diminish the impact of the secondary effects on Singapore and on the region as a whole. This not only applies to natural hazards, but also to disruptions with a manmade element, including pandemics, financial upheavals and threats to cybersecurity.

Taking on this mantle would signal to the rest of the world that, through a combination of leadership, courage, foresight, governance, financial management, education, hard work and determination, the inevitable major disruptions we face need not be existential disasters for humanity. Informing the Southeast Asian region about common threats, and initiating a conversation about how to collectively deal with them, would be a good place for Singapore to start.

ACKNOWLEDGEMENTS

This book would not have been possible without the cooperation of many individuals; some of you have bravely taken to the stage to issue a much needed rallying call, and others have quietly toiled behind the scenes.

Firstly, I would like to thank Para Limes for organising the 'Disrupted Balance—Society At Risk' conference and bringing together our twelve eminent speakers. An institute within Nanyang Technological University, Singapore (NTU), Para Limes is dedicated to the interdisciplinary exploration of complexity. We are grateful to our colleagues at Global Dialogue@NTU for their generosity in hosting the conference.

I would also like to thank Mr Inderjit Singh, chair of the NTU Board of Trustees' Enterprise Committee, for gracing the occasion as our guest of honour. That he would take time out of his busy schedule to attend the talks underscores the importance and relevance of the topics we discussed. Many thanks as well to my conference co-chair Mr Tony Mayer, NTU's research integrity officer, and the heart of Para Limes, Ms Karen Chung, for making sure that the conference ran as well as it did.

I am indebted to the speakers who agreed to deliver their talks in Singapore and contribute a chapter to this book. Their exciting and highly relevant talks were not only deeply researched but also engaging and deeply thought-provoking. I would also like to acknowledge our editorial consultants, Dr Rebecca Tan and colleagues at Wildtype Media Group, for transcribing the talks and editing them into book form. It is my hope that the conference and this book will continue to spur stimulating and necessary discussions on the complex challenges society faces and what can be done about them.

Last but not least, I would like to thank the participants of the conference, whose lively discussions during the coffee breaks and after each of the talks helped to spark many new ideas.

Jan W. Vasbinder

Co-Chair, 'Disrupted Balance—Society At Risk' conference

Exploring Complexity

For four centuries our sciences have progressed by looking at its objects of study in a reductionist manner. In contrast complexity science, that has been evolving during the last 30–40 years, seeks to look at its objects of study from the bottom up, seeing them as systems of interacting elements that form, change, and evolve over time. Complexity therefore is not so much a subject of research as a way of looking at systems. It is inherently interdisciplinary, meaning that it gets its problems from the real non-disciplinary world and its energy and ideas from all fields of science, at the same time affecting each of these fields.

The purpose of this series on complexity science is to provide insights in the development of the science and its applications, the contexts within which it evolved and evolves, the main players in the field and the influence it has on other sciences.